BASIC MATHEMATICS

BASIC MATHEMATICS
Detecting and Correcting
Special Needs

Joyce S. Choate/*Consulting Editor*

BRIAN E. ENRIGHT
**University of North Carolina
at Charlotte**

ALLYN AND BACON
Boston / London / Sydney / Toronto

THE ALLYN AND BACON
DETECTING AND CORRECTING SERIES
Joyce S. Choate, *Consulting Editor*

Copyright © 1989 by Allyn and Bacon
A Division of Simon & Schuster
160 Gould Street
Needham Heights, MA 02194-2310

The figures in Skill Needs 21, 24, 27, 30, 32, 33, 35, 37, 39, 41, 44, 47, and 50 are © 1985 by Curriculum Associates, Inc. Reprinted by permission from the ENRIGHT ™ Computation Series.

Editorial-Production Service: Karen G. Mason
Copyeditor: Susan Freese
Cover Administrator: Linda K. Dickinson
Cover Designer: Susan Slovinsky

Library of Congress Cataloging-in-Publication Data

Enright, Brian.
   Basic mathematics.

   (The Allyn and Bacon detecting and correcting series)
   Includes bibliographies and index.
   1. Mathematics—Study and teaching (Elementary) 2. Handicapped children—Education—Mathematics. 3. Gifted children—Education—Mathematics.
I. Title. III. Series.
 ISBN 0-205-11635-3

Printed in the United States of America

10 9 8 7 6 5 4 3 2 1   93 92 91 90 89 88

Contents

LIST OF FIGURES

Foreword

ABOUT THE *DETECTING AND CORRECTING SERIES*

Basic Mathematics: Detecting and Correcting Special Needs is one of several books in an affordable series that focuses on the classroom needs of special students, both exceptional and nonexceptional, who often require adjusted methods and curricula. The purpose of this book, as well as the others, is to supplement more comprehensive and theoretical treatments of major instructional issues—in this case, teaching mathematics—with practical classroom practices.

The underlying theme of each book in the *Allyn and Bacon Detecting and Correcting Series* is targeted instruction to maximize students' progress in school. Designed for informed teachers and teachers-in-training who are responsible for instructing special students in a variety of settings, these books emphasize the application of theory to everyday classroom concerns. While this approach may not be unique, the format in which both theme and purpose are presented is, in that it enables the reader to quickly translate theory into practical classroom strategies for reaching hard-to-teach students.

Each book begins with an overview of instruction in the given subject, addressing in particular the needs of special students. The groundwork is laid here for both Detection and Correction—observing students' difficulties and then designing an individualized prescriptive program. Remaining chapters are organized into sequentially numbered units, addressing specific skill needs of special students. Each unit follows a consistent two-part format. Detection is addressed first, beginning with a citation of a few significant behaviors and continuing with a discussion of factors such as descriptions and implications. The second part of each unit is Correction, where a number of principles or strategies are offered for modification according to students' learning needs.

This simple, consistent format makes the Detecting and Correcting books accessible and easy to read. Other useful features include: a) the Contents organization designed for quick location of problem skills and behaviors; b) cross-references among units; c) a "Reflections" section ending each part, providing discussion and application activities; and d) an index of general topics and cross-references to related subjects.

Along with the topics of related books—reading, language arts, and classroom behavior—basic mathematics represents a major area in which special students often require special accommodations. Together, these books comprise the first installment of what is envisioned to be a larger series that simplifies teachers' tasks by offering sound and practical classroom procedures for detecting and correcting special needs.

Joyce S. Choate
Consulting Editor

Preface

Basic Mathematics: Detecting and Correcting Special Needs is designed to address the basic mathematics needs of special learners. It is intended for use as a field resource and supplementary text by teachers and prospective teachers who are concerned with improving the mathematics skills of special learners in both regular and special education classroom settings. A practical complement to theoretical texts and teaching wisdom, this book is deliberately brief and concise. The intent is to provide the reader with practical instructional activities and guidelines that are useful when working with special learners in the area of mathematics.

ASSUMPTIONS

In this book, special students include both properly identified exceptional learners and nonexceptional learners who demonstrate significant problems learning basic mathematics when presented in the traditional manner. The basic assumptions underlying the structure and content of this book are these:

- Special students are likely to have special mathematics needs that may vary from those of the typical classroom student who makes average progress.
- Since problem solving is the central focus of the mathematics curriculum, all other mathematics skills should be learned to facilitate the solving of problems.
- Basic mathematics is subdivided into several operational areas, each of which has its own sequential set of skills leading to appropriate problem-solving skills.
- The incorrect answers that students provide to well-constructed assessment items yield important information that points to the procedures for correcting skill needs in mathematics.
- To correct the mathematics problems of special learners, it is appropriate to provide direct instruction in how to accomplish manageable computation tasks and then use those computation skills as tools to solve mathematics problems.
- Although the most appropriate way to teach the concepts of mathematics is through a hands-on/manipulative-based approach, special learners often need a mixture of this approach and a structural step-by-step approach.

- Teaching the use of flowcharts is one simple and effective way to assist special students to master the sequential steps of different mathematics algorithms.
- A corrective program must be based on the analysis of each student's error patterns in basic mathematics.

These assumptions are incorporated into the detection and correction model for identifying the special needs of individual students and then correcting those needs with targeted instruction in basic mathematics.

ORGANIZATION

The Contents is designed to provide an efficient means to locate specific skills quickly. Therefore, once you have reviewed the complete book, you can then use it as a reference in the classroom. For accessibility, categories of special learners and specific mathematics skills are enumerated and discussed as intact units. Although they are treated as separate entities here, in the actual classroom, there is significant overlap. Special learners' behaviors are not always neatly separated and specific students may exhibit several skill needs simultaneously.

The book is divided into six parts. Each begins with an overview and concludes with suggestions for reflecting on the content. "Reflections" are intended for clarification, discussion, extension, and application. In each part, the final "Reflections" item refers to additional resources for further information on mathematics instruction or special learner needs.

The three chapters in Part I describe a framework for special mathematics instruction and the special learner. Chapter 1 briefly explains the validated teaching practices that form the foundation for special instruction in mathematics. These teaching practices provide the framework for the specific corrective strategies discussed in later chapters. The next two chapters present categories of special learners, subdivided into exceptional and nonexceptional learners. Guidelines are offered for developing corrective programs for each group. As mentioned earlier, students in the real world will cross over these neatly drawn lines, and you must take this into account when applying the suggestions provided throughout this book.

Part II describes the various skills of problem solving. The placement of this section before the discussions of computational skills is a philosophical statement regarding the nature of the mathematics curriculum. Although frequently left until last or "stuck in" every so many chapters throughout a given basal text, problem solving is the core of mathematics. Thus, corrective instruction in mathematics can and should revolve around a central focus of problem solving while still correcting specific computation errors.

Readiness skills are the focus of Part III. Parts IV, V, and VI address the various subskills of computation for whole numbers, fractions, and decimals, respectively. These skills are divided into sequential subskills, and the most common error patterns are discussed. Corrective strategies are provided for each.

FEATURES

This book is not intended to replace standard texts but rather to supplement them with a set of practical implementations of sound theory. Included in the strategies is a mixture of concept development and specific subskill correction. Concept-based learning is appropriate for certain skills and certain students, while structural skill learning is most appropriate for other situations. Specifically, if a student has no idea of what addition accomplishes, then he or she needs a program based in concept acquisition. However, if a student seems to understand addition yet demonstrates consistent errors while adding, then a program stressing the mechanics of adding is indicated. Therefore, strategies from both approaches are included for selection according to the situation.

Each skill is organized into DETECTION and CORRECTION sections. In Chapters 4 through 16, all skills are treated in a two-page format, with DETECTION on the left-hand page and CORRECTION on the facing right-hand page. This layout facilitates quick access to information about specific students' needs.

Many of the strategies discussed in this book could be used with any student having difficulty. However, the special student must have this special instruction to progress in mathematics. In particular, this instruction should involve three phases. Each CORRECTION strategy should first be introduced by teacher direction. After that foundation has been laid, instruction can proceed to a paired-learner situation, and, when the student is ready, move to independent study.

ACKNOWLEDGMENTS

These strategies have been developed over several years of field-testing with teachers from across the United States and Canada. I thank those individuals with whom I have worked.

Many of these concepts and strategies were developed during the writing of the *ENRIGHT Diagnostic Inventory of Basic Arithmetic Skills*, *The ENRIGHT Computation Series*, and *Solve...Action Problem Solving*. Indeed, the flowcharts throughout this book come from these sources and are used with permission of the publisher. Therefore, I express my appreciation to Curriculum Associates.

For their feedback and suggestions in refining the text, I am grateful to these field reviewers: Rudolph Cherkauer (State University of New York at Buffalo); George W. Fair (University of Texas at Dallas); and Fred Remer (New York School District #6, Yonkers, New York).

A special thanks to Mylan Jaixen of Allyn and Bacon, who, in addition to being an outstanding managing editor has become a good friend. To my wife and children, who always wonder if Daddy will make this deadline, the answer is "not without their help." Lastly, I would like to thank Joyce Choate, who has supported and encouraged me for several years; it is not the easy times that make enduring friendships but rather the difficult times that test the bonds.

PART I

SPECIAL MATHEMATICS NEEDS
OF SPECIAL LEARNERS

All students occasionally need specific instruction to meet their individual mathematics needs. However, some students require more specialized instruction than others. Part I briefly examines some of the special problems in mathematics experienced by certain categories of students. Whether their difficulties result from learning or experience differences or from poor teaching, many of the students in each classification exhibit similar performance patterns and share special needs in mathematics instruction. Some general guidelines for developing instruction are included both as an overview of corrective mathematics instruction and to describe the needs of special learners. Understanding the similarities as well as the differences among these various special learners will help the teacher deal more effectively with them.

Chapter 1 focuses on special mathematics instruction. It provides an overview of a systematic approach to all corrective instruction based on a diagnostic/prescriptive model. Several general practices are proposed and explained. The ideas proposed in this chapter form the basis for setting up a corrective program with additions or changes to meet individual needs.

Chapter 2 focuses on the mathematics difficulties that students placed in regular and/or remedial education classes frequently exhibit. The problems often described by teachers of these classes are grouped and discussed according to common features. Some may be associated with learning problems, while others are often related to the types of experiences students have or have not had. The extent and nature of the students' experiences regulate the depth and breadth of mathematics comprehension that can occur. Certainly, all students benefit from having numerous learning experiences and ample review of prerequisite skills prior to each mathematics lesson. But special students *must* have this type of enhanced instruction in order to succeed.

Some of the most common mathematics problems of students who have qualified for special education services are treated in Chapter 3. Discussed by classification are the mathematic needs associated with categories of students who have met the criteria for eligibility for special education services. Although these exceptional students may also be taught in the regular or remedial classroom, at least a portion of their instruction is under the direction of special educators.

The boundaries of a single page are used to limit the discussion of each special problem. On each page, you will see listed the observable mathematics behaviors generally associated with the particular learning or experience problem, followed by a cursory description of some typical mathematics performances, implications of the problems to academic progress, and the relationship of the learning or experience problem to mathematics difficulties.

All students occasionally evidence many of the behaviors detected in these special categories of learners. All students who exhibit mathematic difficulties will not fit neatly into one of these categories; nor will every student who matches a classification demonstrate every mathematics problem. Instead, the DETECTION behaviors offer clues to a *possible* pattern of mathematics needs. You will need to synthesize the past and sustained performance of each student and evaluations by appropriate educators with the descriptions, implications, and causation to identify and confirm the exact problems.

You must thoroughly diagnose and analyze the mathematics skills and provide targeted teaching before you can substantiate the applicability of any principle to individual students. These CORRECTIVE PRINCIPLES are intended as preliminary guides for you to select and modify corrective mathematics strategies, such as those presented in later sections.

Note: Due to the similarity of special learners' needs in all the basic skill areas, portions of Part I are adapted from *Reading: Detecting and Correcting Special Needs*, by J. S. Choate and T. .A. Rakes (Boston: Allyn and Bacon, 1989.)

Special mathematics instruction is designed to meet the learning needs of individual special students. Such teaching involves first detecting a student's learning and mathematics skills and then planning instruction that will improve or correct the weaker skills so that they no longer represent a special need.

DETECTION

Central to any targeted instructional plan is the identification of the problem or errors to be corrected. Specific mathematics skills in need of correction may be identified in three ways: 1) through direct testing, using either formal or informal measures; 2) by analysis of daily classroom mathematics work; and 3) by synthesis of both of the above.

Direct Testing

Formal tests of mathematics achievement are a major component of most standardized achievement tests, which are traditionally administered to entire classes of students at the beginning or end of a school year. Whereas these tests are fairly accurate measures of the general achievement of successful students, they can be misleading with regard to special students' abilities.

These tests may give credit for skills students do not have. For example, students who use their fingers to count and find answers in addition are not really adding but using the lower-level skill of counting. Therefore, an achievement test that gives credit only for correct answers may give credit for skills that the student does not have, while not giving credit for partial answers that in fact do demonstrate skill acquisition. Teachers should therefore be very careful in using results from achievement tests. These tests do, however, point out the students who need special help. The use of tests that evaluate answers and determine what the incorrect answer means would be much more useful in determining a solution.

Analysis of Daily Mathematics Work

Unlike direct testing, which requires taking time away from the teaching process, analysis of daily work provides the teacher with ongoing information regarding the effect of instruction on the learning process. This can be extremely powerful information if used to constantly make minor adjustments in the teaching/learning process. A combination of direct testing (taking care not to be misled) and analysis of work on a daily basis will provide the most accurate and helpful information to develop appropriate learning.

CORRECTION

Special students require instruction in most of the same mathematics skills that all other students do. Many of the same instructional procedures that are appropriate for teaching mathematics to all students are equally appropriate for teaching mathematics to the special learner. However, variations of the accepted methods render them more effective for correcting the mathematics needs of special students.

Controversy abounds regarding the so-called correct approach to teaching mathematics. The conventional wisdom proposes the exclusive use of manipulatives (concrete objects) so that students might understand the concepts of mathematics. This approach is clearly the approach of first choice; however, not all students learn best this way. Some students have significant trouble making the transition from understanding to doing. For these students, a second, more structurally oriented approach is needed to facilitate transfer from concrete to abstract.

The ideal approach would be to combine these two supposed opposites, employing the strengths of both. Doing so could produce far greater gains by special students.

Teach Diagnostically

Of the teaching practices that are particularly important for working with special learners, diagnostic teaching is the most effective. Diagnostic teaching is not so much a specific instructional approach as it is a management system. It involves three basic steps: 1) Assess to locate the student's strengths and weaknesses; 2) develop and deliver a plan directed at these weaknesses; and 3) continuously monitor the plan to determine which elements of the plan are proving most helpful.

This model helps the teacher merge the two elements of instruction into one plan: The skills the student needs to work on are developed through the most effective manner in which that student learns.

Use Manipulatives to Develop Basic Concepts

Manipulatives are concrete objects that students handle and move about in order to experience mathematics and to realize the relationship of numbers to real-life situations. Students can build various groups of objects, learning concepts of size, greater or less than, and basic operations. It should be noted that students will count when they work with objects and that this process must later be transferred to the abstract level.

Frequently, students stop at the counting level, and that can slow future progress. For example, a student might have two groups of blocks on a table in front of him or her. The task is to find out how many blocks are in each group and how many there are in all. The student counts the blocks in group A (3) and then the blocks in group B (2). Students who rely too heavily on counting will often go back and count each block again to determine the total number.

The student should recognize that there were three blocks in group A and begin counting the total from that point (i.e., group B would include blocks 4 and 5). This may seem very basic, but many special learners never leave this level of functioning because the transfer from the concrete level to the abstract was never directly taught. Therefore, it is essential to stress the transition period, even when planning for the concrete level of instruction, to prohibit the fixation at that level.

Use Real-Life Experiences to Teach Mathematics

Mathematics is an area that can easily be built around students' lives and experiences. Almost every student has accompanied a parent to a store or has gone up a flight of stairs or has collected some vast amount of childhood treasures (rocks, bottlecaps, or buckeyes). All of these and countless other experiences can be used to make problem solving and even computation come alive. Consider the students' perspective: Instead of just adding the same old numbers, those numbers represent quantities of items you are collecting. Or instead of working on a word problem about an experience Dick, Jane, and Spot had, the problem is about a trip the student has just taken. This approach can be expanded to the whole class as well. When a student returns from a trip,   he shares some of his discoveries. The problems for mathematics that day are built around the trip. The entire class joins in.

This does not imply that commercially prepared materials are not good. Rather, it means that the teacher should use those materials that his or her students can relate to and that good teaching practices find ways of making that connection.

Apply Principles of Behaviorism

Principles of behaviorism have been successfully applied to improve academic achievement of special students. Behaviorism is based on the premise that behavior that is rewarded is more likely to recur. Thus, two major factors must be considered when planning a program for teaching mathematics: 1) Define the mathematics behavior to be changed through observation and diagnosis, and 2) design a program to systematically reinforce desired responses.

Factor one deals with identifying what must be learned. It further requires the teacher to break that skill down into manageable steps that can be observed and reached in a reasonable amount of time. Step two is equally individual in nature. First, the teacher must find what indeed is reinforcing to a given student. Next, the teacher must determine how often a given student must be reinforced.

Use Flowcharts to Develop Proper Algorithms

Flowcharts are visually oriented presentations of the step-by-step sequence of an algorithm. Although they are clearly not needed by many successful learners in mathematics, they provide especially strong assistance to learners with sequential learning difficulties. Students with sequential learning problems often perform the steps of an algorithm out of sequence or leave out steps altogether. These same students have difficulty sequencing subskills in many other content areas.

Flowcharts serve these learners as roadmaps through the skill sequence and can be dispensed with once that sequence has been mastered. They are particularly useful in areas in which subskills develop by adding additional steps. Flowcharts are based on the product of task analysis. In fact, they can only be developed after a task analysis has been completed. This is a good example of merging a particular skill deficit with a particular method of teaching.

Extend Problem-Solving Skills throughout the Curriculum

The basic skills needed for problem solving in mathematics are similar to the skills needed for problem solving throughout the curriculum. As students learn each skill, have them find examples of how that skill could be used in other curriculum areas. This is often done in student pairs, which can reduce stress and increase brainstorming potential. After the students have found applications in other subjects, have them apply the skills to areas of interest outside school, such as sports or hobbies. This activity not only boosts interest but enables learners to synthesize and incorporate problem-solving skills into their learning system.

Teach Students to Attend to Detail

Mathematics is an area in which detail matters. Surely, in reading, the facts are important, but in mathematics, they are essential. Imagine the problem that requires a student to work with 20 sets of three objects, but the student mistakenly works with two sets of three objects.

On the one hand, detail is not enough, for without the concept, the student won't know what to do with the detail. However, on the other hand, concept is also not enough, and special students who are not taught to attend to and organize details will grow up to be adults whose checkbooks are always out of balance.

Teach Self-Monitoring

In order to become successful and independent learners, students must be able to monitor their own work. They will need to address two variables as they develop self-monitoring skills: quality of work and quantity of work.

Quality of work refers to correctness: Are the answers accurate? Does the answer match the question? To determine if the answer matches the question, you must first have a clear understanding of what the question is. This provides the learner with something to compare the answer to.

Quantity of work is also important for special learners to monitor. For these learners to compete successfully in the real world of the classroom, they must complete significant amounts of work. After a specific skill has been learned, the student should monitor himself or herself to track increases in the speed of work and consequently the amount of work completed correctly in a given time. This self-monitoring can be done simply, using a table on which the student records the work attempted and accurately completed for each time period. In addition to building independent skills, this process should begin to provide the student with positive feedback.

Develop Paired Problem-Solving Skills

Unfortunately, teachers cannot be available to all students at all times. When students work in pairs, they can often get through the difficult areas of problems and both grow in the process. Students who cannot solve a problem on their own frequently benefit from working with a peer.

Student pairs start by independently reviewing the problem and then discussing it. They each formulate a plan and then compare the plans to see if one might work better. It is important that the two solutions be derived independently and then compared. If the solutions are the same, the chances are good that they are correct. If the solutions are different, the student pair must try to determine why. If they are unable to resolve the dilemma, only then is the teacher brought in to help determine an acceptable solution. This process not only frees the teacher, providing time to spend with students experiencing serious problems, but also causes the students to become more active in the learning process.

Check All Materials for Readability

Teachers frequently assume that mathematics books are controlled for reading level much the same way that reading basals are. This is not always the case. It is necessary to check the readability and useability of all student materials to determine what is considered *on level*. If students are experiencing serious problems in reading in addition to mathematics, providing on-level work in on-level material becomes even more complicated. However, providing on-level tasks in frustration-level material will prove unsuccessful.

Almost every introductory reading text offers several readability formulas; this will provide a beginning. However, the teacher needs to check the vocabulary for familiarity of sight and meaning. Compare words in the text that students may be unsure of to a common word list; the Dolch and Harris-Jacobson are two common lists. Key direction words are extremely important, and if need be, can be learned using flashcards. Although this preteaching activity might seem cumbersome, it will help avoid needless student frustration.

Teach for Mastery

Mastery of basic skills is essential for special learners to acquire more advanced skills. Many times, special learners' objectives are considered to be complete when the student scores correctly one time. This is unfortunate because these learners almost always require many successful repetitions before they are able to move on.

In addition, special learners need periodic review even more than their nonspecial counterparts. This "spiraling" back throughout the year can help special learners truly master the skills. With special learners, this review may need to be done as often as every few days. Because mathematics involves such a sequential building of skills, the spiraling can fit very naturally into every lesson. For example, the last five problems in any assignment might be reserved for reviewing previously acquired skills.

SUMMARY

These instructional practices are incorporated into the corrective strategies and techniques throughout the book. In many cases, because of the similarities of the skills and/or the error patterns, strategies are repeated.

SELECTED REFERENCES

Algozzine, B., & Maheady, L. (Guest Eds.). (1986). In search of excellence: Instruction that works in special education classrooms [Special issue]. *Exceptional Children 52*(6), 481–608.

Choate, J. S., Bennett, T. Z., Enright, B. E., Miller, L. J., Poteet, J. A., & Rakes, J. A. (1987). *Assessing and programming basic curriculum skills.* Boston: Allyn and Bacon.

Enright, B. E. (1983). *ENRIGHT diagnostic inventory of basic arithmetic skills.* N. Billerica, MA: Curriculum Associates.

Enright, B. E. (1986). *ENRIGHT computation series* (Books A–M). N. Billerica, MA: Curriculum Associates.

Enright, B. E. (1987). *SOLVE: Action problem solving* (Books I–III). N. Billerica, MA: Curriculum Associates.

Gelzheiser, L. M., & Sheperd, M. J. (Guest Eds.). (1986). Competence and instruction: Contributions from cognitive psychology [Special issue]. *Exceptional Children 53*(2), 97–192.

National Council of Teachers of Mathematics, Commission on Standards for School Mathematics. (1987). *Curriculum and evaluation standards for school mathematics* (working draft). Reston, VA: NCTM.

Squires, D. A., Huitt, W. G., & Segars, J. K. (1984). *Effective schools and classrooms: A research-based perspective.* Alexandria, VA: Association for Supervision and Curriculum Development.

Wang, M. C., Reynolds, M. C., & Walberg, H. J. (Eds.). (1987). *Handbook of special education: Research and practice* (Vols. 1–2). New York: Pergamon Press.

CUMULATIVE-DEFICIT LEARNERS

DETECTION These students may:

- Demonstrate a pattern of steadily decreasing performance in all subjects
- Perform only basic computation skills
- Have difficulty applying algorithm skills
- Answer the easy questions but not the hard ones
- Appear to learn at a slower rate than most peers

Description. Students with cumulative-deficit problems are most often described by teachers as the ones who "got behind and can't catch up." These students simply did not master the prerequisite mathematics skills before the curriculum moved on ahead of them. When too many parts or skills are missing, the total mathematics performance suffers.

Causation. Cumulative deficits occur when the pace of the curriculum is faster than the rate of the students' mathematics progress. When students lack prerequisite skills, mathematics deficits are compounded as new skills are introduced and the difficulty level of the mathematics text increases. Frequent or lengthy absences, a change in schools or mathematics series, and fragmented instruction caused by teacher error or absence interrupt program and skill continuity.

Implications. A cumulative deficit in mathematics skills eventually interferes with performance in many subject areas. Students who struggle with basic computation will be unable to attend to problem solving in mathematics and other subjects. Classwork is often incomplete or performed incorrectly because of inadequate understanding of concepts and/or mechanics.

CORRECTIVE PRINCIPLES Use these guidelines to plan and modify instruction.

1. Backtrack and teach to the point of deficiency in each skill sequence.
2. Temporarily slow instructional pace to permit reteaching of skills.
3. Review and reteach prerequisite skills before each lesson.
4. Provide a tutor for supplementary instruction in appropriate prerequisites.
5. When teaching new algorithms, use flowcharts and stress the relationship to previous skills.
6. Use paired problem solving to introduce new skills.

DISADVANTAGED LEARNERS

DETECTION These students may:

- Display a restricted vocabulary in mathematics
- Have difficulty changing from one computation skill to another
- Often comprehend surface meanings only in word problems
- Exhibit disinterest in mathematics
- Perform below capacity level

Description. Disadvantaged students are those whose deficient home environment contributes to mathematics problems. Disadvantaged learners often appear capable of higher mathematics performance than they demonstrate, but many do not display a great amount of interest in learning or improving mathematics skills. Often, these students experience difficulty interpreting word problems and questions being asked. Because of their restricted vocabulary and experiences, many of these students' problem-solving difficulties resemble those of the language-different learners.

Causation. Regardless of economic level, these students have been deprived of the types of stimulation that contribute to mathematics success. Although most disadvantaged learners are from low-income homes, a few come from affluent families who, for some reason, overlook the children's needs for intellectual and quantitative stimulation. These homes are not likely to provide meaningful practice, enrichment, and reinforcement to supplement the school's mathematics program. Disadvantaged learners tend to begin school unprepared to receive formal mathematics instruction; without targeted instruction in quantitative skills and a concerted effort to extend experiences, these students continue to lag behind their age peers in learning mathematics.

Implications. Accustomed to functioning in an unstimulating environment, these students often do not know how or even particularly want to learn. They must be motivated and taught goal orientation as well as specific learning strategies. These needs, coupled with narrow experiences and language skills, often contribute to poor student performance in mathematics as well as in other subjects.

CORRECTIVE PRINCIPLES Use these guidelines to plan and modify instruction.

1. Use manipulatives to enhance concept development.
2. Build concrete and vicarious experiences prior to each mathematics lesson.
3. Extend the readiness activities for each lesson, stressing quantitative meanings.
4. Stress the relationships between sets of objects and numbers.
5. Directly teach learning strategies and specific strategies for each skill.
6. Establish manageable mathematics goals and reinforce accomplishments.

LANGUAGE-DIFFERENT LEARNERS

DETECTION These students may:

- Often rearrange or substitute parts of mathematics algorithms
- Have difficulty understanding special mathematics terms
- Frequently comprehend only surface meanings in word problems
- Perform better when reading and working silently

Description. Students whose primary language and/or culture differ from the predominant school culture may or may not display mathematics difficulties. Although computation skills may not necessarily suffer, the language difficulty often interferes with understanding story problems. Students' performance tends somewhat to resemble that of disadvantaged and language-disabled students. Silent work is often less difficult for language-different students than is class discussion about mathematics.

Causation. Many of these students think in their primary language and then translate into Standard American English. This means that some of the idiosyncracies of the first language often creep into the translation of story problems, producing mathematics errors consistent with the student's dialect or language. The translation process also slows the independent performance of mathematics lessons because written directions must be interpreted.

Implications. Language differences complicate the learner's task in mathematics as well as in other subjects. In all subjects, it is often necessary to use pictures and other visual cues to explain and translate concepts. The cultural differences that accompany language differences often mean experiential backgrounds that differ from those of the school majority. Common experiences must be established or explained prior to mathematics lessons, particularly problem-solving lessons. Be sure to isolate and teach technical vocabulary early in the learning process.

CORRECTIVE PRINCIPLES Use these guidelines to plan and modify instruction.

1. Use pictures and other visual cues to explain and translate concepts.
2. For initial instruction in story problems, use diagrams in place of words.
3. Build concrete and vicarious experiences prior to each mathematics lesson.
4. Concentrate on silent comprehension, not pronunciation, with word problems, allowing students extra time to accomplish mathematics assignments.
5. Try a mathematics hands-on approach to bridge the gap between students' oral language and mathematics.

SLOW LEARNERS

DETECTION These students may:

- Make steady but slow progress in mathematics
- Require more practice and repetition than most age peers
- Have a narrow speaking and listening vocabulary in mathematics
- Take longer than age peers to compute answers
- Need to reread problems to answer questions

Description. These students are often described as those who "always lag behind the rest of the class." Although the skill sequence of the regular mathematics curriculum is appropriate for these students, they take longer to master each task. A slightly slower pace and additional instruction and reinforcement are required for them to acquire the important mathematics skills.

Causation. These students need not experience mathematics difficulties if the pace of the mathematics curriculum is adjusted to their slower rate of learning. However, the typical mathematics problems occur when the pace of the curriculum leaves them behind. Mathematics deficits begin to accumulate as these students face new and more difficult tasks before mastering the prerequisite skills. As with the students described as exhibiting cumulative deficits, factors such as narrow technical vocabularies, limited experiential backgrounds, and limited mathematics practice at home further contribute to the problem.

Implications. Although slow-learning students may not advance a full mathematics grade level each school year, many can master the most basic of the mathematics skills. Typically, their slow learning rate is not confined to mathematics but is reflected in their overall academic performance. They will require supplementary instruction in comprehension to learn problem-solving skills.

CORRECTIVE PRINCIPLES Use these guidelines to plan and modify instruction.

1. Slow the pace of mathematics instruction to match the students' learning rate.
2. Teach to the point of deficiency in each mathematics skill sequence.
3. Review and reteach prerequisite skills before each lesson.
4. Provide extended readiness and practice activities for each lesson.
5. Stress the use of the most reliable computation strategies.
6. Directly teach word meanings for technical vocabulary.
7. Shorten mathematics assignments to manageable units.

TEACHER-DISABLED LEARNERS

DETECTION These students may:

- Evidence mathematics skill gaps
- Apply mathematics skills inappropriately
- Either try very hard or not at all
- Have difficulty comprehending the question in problem solving
- Compute at a slow rate

Description. Teacher-disabled learners are those who have been taught incorrectly or those about whom teachers say, "They have just never been taught the basic mathematics skills." The mathematics behaviors of these students are often of one of two extremes: Either the student obviously exerts a tremendous amount of effort, or he or she appears to have given up in defeat. Many of these students appear to know only one strategy for mathematics, often concentrating on counting at the expense of comprehension. Some laboriously count out each column, never learning the relationship between the numbers, and then quite naturally fail in the attempt to regroup or use higher-level skills.

Causation. The reasons for learners becoming teacher disabled are varied and are often suspected after the fact. The prime offenders are those nonteachers who put the books and workbooks with the children, assuming mathematics will happen, and those who teach the basal mathbook, not the students, insisting that the entire class advance together according to a preplanned agenda. Other contributing factors include frequent teacher absences, changes in basal mathematics programs, rigid teachers, those who in effect teach students that they can't do math, and teacher/student personality conflicts. Some teachers overemphasize one skill at the expense of others, producing learners who use only one strategy that is often inappropriate. Although these are initially teacher, not student, problems, they eventually result in students' mathematics problems.

Implications. Students who conscientiously follow the instructions of their teachers without success soon lose faith in themselves and their education. Without intervention, they are destined to experience difficulty in mathematics as well as across the academic curriculum. Slowed by ineffective or inappropriate mathematics strategies, the slow performance rate of many of these students may be reflected in incomplete assignments.

CORRECTIVE PRINCIPLES Use these guidelines to plan and modify instruction.

1. Backtrack and teach to the point of deficiency in each mathematics skill sequence.
2. Emphasize understanding, using manipulatives.
3. Review and reteach prerequisite skills before each lesson.
4. Build the basic facts through fun drills.
5. Carefully structure mathematics experiences to assure students' success.

UNDERACHIEVING LEARNERS

DETECTION These students may:

- Perform mathematics tasks below the level of teacher expectation
- Perform inconsistently and be distractible
- Solve problems much better when the topic is of high interest
- Appear to dislike and avoid the rigors of academic work
- Often comprehend surface meanings only in word problems

Description. Underachievers usually exhibit similar performances across school subjects. They are the ones about whom teachers say, "He could do it if he wanted to" or "She can do it when you make her." Much like reluctant learners, they occasionally work voluntarily when a topic appeals to them and even show sudden bursts of skill gains, but the enthusiasm soon disappears. Many of these students appear to be somewhat easily distracted in the classroom.

Causation. The causes of underachievement are many and varied. Lack of motivation is often a primary factor. This may result from sociocultural values in conflict with those of the school, lack of reinforcement for academic achievement, negative role models, and inability to see the personal relevance of achievement. Teacher pressure may only compound the problem. Limited experiences, language differences, and inadequate educational training can also contribute to underachievement. Students who are easily distracted by peers cannot easily concentrate while doing mathematics. Also consider the possibility of specific learning differences such as those discussed in Chapter 3.

Implications. Students who compute well below their capacity do so at the cost of developing the skills prerequisite to future progress in mathematics. They often appear to exert little effort, and their understanding may be limited to surface meanings. Teachers are often impatient with these students, but when they try to make underachievers do the work, they may cause the students to dislike mathematics.

CORRECTIVE PRINCIPLES Use these guidelines to plan and modify instruction.

1. Focus on activities that emphasize the usefulness of mathematics.
2. Clearly state the purpose of each mathematics activity.
3. Allow for purposeful movement, but limit distractions while doing mathematics.
4. Teach students computation skills they can use in their environment.
5. Until students are motivated, deemphasize skill and drill and emphasize mathematics for survival.
6. Determine students' interests and then find appropriate mathematics materials.

CROSS-CATEGORICAL HANDICAPPED LEARNERS

DETECTION Properly identified students may:

- Exhibit difficulty mastering certain mathematics skills
- Perform work below the level of age peers
- Perform mathematics assignments at a slow rate
- Evidence some of the mathematics problems cited by exceptional categories on the pages that follow

Description. Some school systems provide special education services to students in a cross-categorical or generic fashion, grouping students according to the severity of the handicap. Thus, students who are classified as mildly to moderately handicapped are those whose disability is one of a mild to moderate degree and who evidence many of the learning characteristics of the categorical exceptionalities. Although such students are eligible for special education instruction, many of them spend the greater portion if not all of the school day in a regular education classroom. A large percentage of these students perform mathematics tasks below the level of their age peers and experience difficulty mastering some mathematics skills; many also exhibit some of the mathematics behaviors more typical of one or more categorical exceptionalities.

Causation. These students qualify for special education services on the basis of their unique learning needs for adjusted instruction and curricula. These same unique needs often interfere with the orderly acquisition of the skills required for the complex task of mathematics. Among the characteristics shared by many of these students are short attention span, distractibility, need for success and encouragement, specific reading disabilities, limited experiences, and the need for individualized specialized instruction in mathematics. Some of these learning characteristics appear to complicate mathematics tasks, while others directly interfere. For specific examples of the interactions of learning characteristics and mathematics performance, refer to the treatments of the mathematics needs of each of the exceptional categories presented in the pages that follow.

Implications. Although cross-categorical students share many common mathematics needs, differences in their learning characteristics produce varied mathematics profiles. Yet most of their mathematics patterns indicate both general and specific mathematics problems. The difficulties they experience in mastering mathematics tasks are often reflected in difficulties across the academic curricula.

CORRECTIVE PRINCIPLES

The applicability of numerous common principles of corrective instruction to mildly and moderately handicapped students in several exceptional categories provided much of the impetus for cross-categorical or generic groupings. Many of these principles are equally appropriate for teaching mathematics to nonhandicapped students. They recommend structuring the mathematics environment, building appropriate experiences and attitudes, and adjusting the content and style of instruction to the learner's needs. These principles are presented as guidelines for planning and modifying mathematics instruction for mildly and moderately handicapped students.

1. Build and recall experiences to prepare students for each mathematics task.
2. Determine learning strengths and then plan mathematics approaches accordingly.
3. Vary the stimulus/response pattern according to students' needs.
4. Supplement visual information with auditory cues and auditory information with visual cues.
5. Present short and varied mathematics tasks.
6. Keep classroom distractions to a minimum.
7. Plan mathematics experiences that assure students' success.
8. Introduce new mathematics skills slowly and as students are ready.
9. Consider students' interests as well as level of difficulty when selecting mathematics material.
10. Practice mathematics problem-solving skills using interesting problems.
11. Provide opportunities for purposeful movement before, during, and after mathematics lessons.
12. Systematically reinforce appropriate mathematics behaviors.
13. Include review material or tasks as the major portion of each mathematics activity.
14. Provide periodic review and reinforcement activities.
15. Build the necessary technical vocabulary slowly.
16. Provide specific instruction and experiences to broaden vocabulary.
17. Use manipulative activities to introduce and reinforce difficult concepts.
18. Use focused activities to teach comprehension skills and their relationship to problem solving.
19. Emphasize the usefulness and relevance of mathematics.
20. Clearly state the purpose for presenting each task.

BEHAVIOR-DISORDERED LEARNERS

DETECTION Properly identified students may:

- Work below the theoretical expectancy level
- Have difficulty attending to mathematics tasks
- Compute either slowly or very quickly
- Have difficulty beginning or completing math tasks
- Become agitated when difficult problems or texts are encountered

Description. The nature of mathematics difficulty, if any, is dependent in part upon how the behavior is manifested. When mathematics performance is adversely affected, it is often reflected in unsatisfactory mathematics rates, inaccuracy, and poor sustained and/or group performance, as well as an escalation of negative behavior when mathematics tasks become difficult.

Causation. The mathematics problems experienced by behavior-disordered students are often a logical outgrowth of their behavior problems. Distractibility interferes with on-task behavior. Insecurity in unstructured or changed mathematics situations impedes performance. Overcautiousness and distrust create careful but slow computers. Impulsivity results in fast work with decreased accuracy. Low frustration tolerance causes defeat when facing difficult tasks. Disruptive behavior interferes with performance of self and others. Group activities overstimulate some students and intimidate others.

Implications. Behavior-disordered students do not necessarily exhibit mathematics difficulties. In fact, some become very proficient mathematicians. However, many are reluctant to participate in group activities and prefer to engage in mathematics as a solitary task; these students may satisfactorily perform independent assignments but not interactive assignments. Certain word-problem themes are aversive to some students and decrease performance. The major threat to academic performance is the interruption in sequenced skill mastery, which may eventually result in a cumulative mathematics deficit.

CORRECTIVE PRINCIPLES Use these guidelines to plan and modify instruction.

1. Consider the student's personal needs when planning a mathematics program.
2. Involve students in the planning of mathematics activities.
3. Plan highly structured programs, using a consistent stimulus/response format, mathematics routine, and positive reinforcement.
4. Adjust for students' attention spans and frustration tolerance levels.
5. Contract with students for specific skills and then chart progress.
6. Stress the use of real-life computation skills.
7. Use the students' own stories as a source of math word problems.

HEARING-IMPAIRED LEARNERS

DETECTION Properly identified students may:

- Evidence difficulty solving word problems
- Master math skills slowly
- Have a limited vocabulary
- Have difficulty explaining answers

Description. Students who are hearing impaired often evidence many of the same difficulties as those with general language disorders and some of the problems of those with speech disorders. Their command of language is often inversely proportional to the degree of hearing loss. They tend to require extended instruction and practice to master word-problem skills.

Causation. Denied the wealth of language experience available to their hearing peers, these students tend to struggle with number/word meaning and the manipulation of words and ideas. Hearing-impaired students must strain to hear, then draw from an often sparse vocabulary to express ideas, monitor their speech, and again strain to hear another response; thus, many such students hesitate to interact verbally. This further limits their language skills as well as their demonstration of mathematics achievement. They must have shortened lessons because they tire easily due to the amount of effort expended; yet they require extended instruction to master problem-solving tasks. Thus, they may progress slowly, at a rate that often does not permit them to maintain the curricular pace of peers. This slow pace can easily lead to a cumulative mathematics deficit.

Implications. Like students with language disorders, these students tend to experience particular difficulty in learning and improving problem-solving skills. They often miss much of the incidental learning available to their hearing peers. Their narrow vocabularies and difficulties understanding subtle meanings interfere not only with mathematics progress but also with progress in other academic areas.

CORRECTIVE PRINCIPLES Use these guidelines to plan and modify instruction.

1. Use manipulatives to demonstrate concepts prior to learning the algorithms.
2. Seat students where they can best see and hear the teacher.
3. Present visual cues for each mathematics task.
4. Alternate auditory and visual mathematics tasks.
5. Extend the readiness and practice activities for each mathematics skill.
6. Provide corrective feedback in a positive manner.
7. Offer extensive training in the meanings of technical words.
8. Carefully build the experiential framework before mathematics lessons.

LANGUAGE-DISABLED LEARNERS

DETECTION Properly identified students may:

- Evidence a specific comprehension difficulty in problem solving
- Have a limited and/or inappropriate technical vocabulary
- Pronounce words but not know their meanings
- Have difficulty explaining answers

Description. Understanding the meanings of words, phrases, and passages can be particularly difficult for language-disabled students. Although they may correctly pronounce words with ease, they may not interpret the meanings of the words or the word problems. In conversation, they often speak haltingly, as though searching for the correct words and trying to put them together. Some have apparent difficulty understanding class discussions, verbal explanations, and examples.

Causation. The mathematics problems of language-disabled students are generally ones of meaning. Difficulties receiving and interpreting language may interfere with understanding the meanings of the words and ideas in word problems or oral directions and discussions. These students' assimilation of text is hindered by their inability to skillfully manipulate words and ideas. Much of the incidental learning that occurs from various language interactions is also denied them. Difficulties expressing thoughts, whether verbally or in writing, may interfere with their demonstrating what they do know and understand.

Implications. Since mathematics is a type of language skill, these students tend to experience particular difficulty in learning and improving problem solving. Some students may actually understand text but appear unable to formulate the words to express their knowledge. Thus, when asked to select correct answers, as in a multiple-choice activity, they can demonstrate their understanding, but when asked to express their knowledge in short-answer or essay form, they cannot. Often classroom directions and oral explanations are not fully understood, leading to misinterpretations that further compound the problem.

CORRECTIVE PRINCIPLES Use these guidelines to plan and modify instruction.

1. Provide intensive instruction in the meanings of technical words.
2. Use manipulatives to demonstrate concepts.
3. Supplement instruction with visual aids.
4. Evaluate word-problem performance by having students select the correct answer.
5. Build a background of concrete experiences for each problem.
6. Teach problem-solving skills first as a listening experience.

LEARNING-DISABLED LEARNERS

DETECTION Properly identified students may:

- Evidence inconsistent and uneven mathematics performance
- Exhibit mathematics skill gaps
- Be easily distracted from mathematics tasks
- Have difficulty remembering words or technical word meanings
- Forget facts
- Experience difficulty with certain problem-solving subskills

Description. Learning-disabled students often experience significant difficulty in mathematics. The exact manifestation of their mathematics difficulties, however, varies widely. Many are easily distracted from mathematics tasks. Some find fact memorization particularly difficult; others find an algorithmic approach frustrating. The most extreme cases find neither approach effective. The patterns of problem solving also vary; some students may understand the global ideas in problems but ignore details, while others recall facts but miss the interpretations and general messages. Some students comprehend much more when reading orally, while others must read the problems silently.

Causation. Primarily a heterogeneous group, learning-disabled students typically exhibit a disparity between mathematics achievement and theoretical mathematics capacity. Many of their learning characteristics compound their mathematics difficulties. A tendency toward hyperactivity, distractibility, and short attention span interrupts on-task mathematics behaviors. Fluctuating performances result in frustration for students and teachers. The apparent processing difficulties further confuse these students.

Implications. The mathematics problems of the learning disabled are varied and often complex. The same inconsistencies evident in their performance of mathematics tasks appear across curricular areas, and many of the instructional strategies that increase mathematics progress also improve performance in other subjects. Adjusting the stimulus/response requirements of mathematics tasks to these students' needs tends to increase their mathematics performance. Some specialists suggest teaching the learning disabled according to their preferred learning style, either auditory, visual, or combined tactile and kinesthetic.

CORRECTIVE PRINCIPLES Use these guidelines to plan and modify instruction.

1. Determine the student's preferred learning style and teach to it.
2. Prepare brief mathematics activities, and then vary them.
3. Teach students strategies to compensate for specific learning weaknesses.
4. Provide a nondistracting mathematics environment.
5. Teach the facts using manipulatives for understanding.
6. Build flowcharts for each computation skill to show the routine.

MENTALLY RETARDED LEARNERS

DETECTION Properly identified students may:

- Make very slow progress in mathematics
- Have a limited mathematics vocabulary
- Forget number facts without numerous repetitions
- Recall details but not sort out extraneous data
- Have significant difficulty solving word problems

Description. The limited intellectual capacity of mentally retarded students similarly limits their mathematics potential. Their rate of progress is noticeably slow. They require extended time and instruction to master mathematics tasks, although the sequence of skill mastery is that of the regular math curriculum. While these students may master the rudiments of computation, problem solving is mastered slowly and laboriously.

Causation. Most of the mathematics problems experienced by mentally retarded students are a reflection of their limited language and thinking skills. Due to their narrow language skills, these students often have difficulty understanding technical word meanings and nontechnical terms with special meanings. The reading skills and the higher levels of cognitive processing required for problem solving are particularly difficult for these students. They also tend to overlook subtle meanings and peripheral and incidental knowledge.

Implications. Because of their slow rate of learning, these students should not be expected to advance a full mathematics grade level each school year. This slow progress is not confined to mathematics but permeates their academic performance. They may require supplementary instruction in recognizing and understanding terms essential to following written directions as well as to consumer mathematics skills.

CORRECTIVE PRINCIPLES Use these guidelines to plan and modify instruction.

1. Teach mathematics very slowly, reviewing frequently.
2. Extend the readiness activities for each mathematics lesson, building prerequisite skills and concepts.
3. Use manipulatives to teach and reinforce concepts.
4. Often repeat instruction and practice for each mathematics task.
5. Follow the skill sequence of the regular mathematics curriculum.
6. Consider a very structured, step-by-step approach with little variance.
7. Directly teach the meanings of the words in the context of problems.
8. Use real-life situations for problem solving.

PHYSICALLY/MEDICALLY HANDICAPPED LEARNERS

DETECTION Properly identified students may:

- Perform mathematics assignments at a slow rate
- Evidence inconsistent mathematics performance
- Exhibit mathematics skill gaps
- Have a limited mathematics vocabulary
- Experience difficulty comprehending and interpreting text for problem solving

Description. Physically/medically handicapped students have physical or health problems that so limit their classroom performance that they are eligible for special education services. When mathematics is adversely affected, it is often in the areas of word meanings, performance rate, and specific gaps in the sequence of mathematics skills.

Causation. Physically/medically handicapped students are frequently absent from school because of health complications; mathematics skill gaps often result. The physical adjustments some learners must make to accomplish mathematics and related tasks, particularly written assignments, often require such strain and effort that the learners perform slowly. They also tire quickly, and many must have mathematics lessons of short duration. Physical and health limitations similarly limit the range of experiences available to some students; this lack of experience may be reflected in shallow comprehension and narrow vocabularies. Mathematics performance may fluctuate in direct relation to the condition of the student's health. Without environmental adjustments and focused instruction, a cumulative mathematics deficit may occur.

Implications. The nature of the physical or medical handicap determines the degree, if any, of interference with mathematics performance. Students who are mobile, attend school regularly, and feel well most of the time do not necessarily experience mathematics difficulty resulting from the handicap. Alterations in the classroom environment—including changes in the physical structure of the classroom itself, the use of mechanical devices, and alternate stimulus/response patterns—will often compensate for physical deficits. The need for additional experiences and vocabulary expansion may cause some of these students to progress slowly in mathematics.

CORRECTIVE PRINCIPLES Use these guidelines to plan and modify instruction.

1. Broaden the experiential repertoire where needed.
2. Adjust the physical environment to accommodate students' needs.
3. Present short and varied mathematics tasks to avoid fatigue.
4. Provide intensive instruction and experiences to broaden vocabulary.
5. Develop a manipulative-based program that these students can handle easily.

SPEECH-DISORDERED LEARNERS

DETECTION Properly identified students may:

- Be reluctant to recite math problems aloud
- Have difficulty discriminating certain sounds
- Not speak fluently or with appropriate voice
- Seldom volunteer oral answers to questions
- Comprehend better when working silently than orally

Description.   Speech disorders are of three types: articulation, fluency, and/or voice. Each generally distorts the student's oral production of text. These students may be reluctant to orally work math or answer aloud. For some, certain sounds are particularly difficult to discriminate and/or produce. Silent comprehension is often easier for these students, as they are freed from the task of pronunciation and can attend to the meaning of the problem.

Causation.   The particular type of speech disorder generally determines the nature and degree of interference with oral answering. Problems of articulation result in consistent mispronunciations of specific sounds and words. Fluency difficulties cause the repetition or prolonging of certain sounds. Voice disorders result in inappropriate pitch, intensity, or quality of voice and cause the student to tire easily. The speech effort and distortions often interfere with comprehension of text when reading word problems aloud.

Implications.   Speech disorders need not necessarily interfere with mathematics progress. Although oral recitation may suffer as a consequence of the great effort and concentration focused on speech production and the actual speech distortions themselves, silent work can be substituted for oral tasks. Instruction in discriminating and producing particularly difficult number words should be coordinated by the speech therapist. The most obvious interference with mathematics performance is that students are reluctant to answer aloud voluntarily, thereby limiting oral interaction and the display of skills and knowledge. Teachers and peers are often embarrassed for these students and thus answer for them or don't give them ample opportunities to demonstrate their mathematics achievements.

CORRECTIVE PRINCIPLES Use these guidelines to plan and modify instruction.

1. Develop choral drills for math-fact mastery.
2. Encourage oral recitation and answering in a supportive environment.
3. Provide corrective feedback in a positive manner.
4. Provide ample opportunities to practice fluent speech.
5. Offer extended practice for discriminating the sounds of target numbers.
6. Permit silent reading of word problems.

VISUALLY IMPAIRED LEARNERS

DETECTION Properly identified students may:

- Compute at a slow rate
- Lose their place on the page
- Hold texts at an unusual distance
- Often confuse similar numbers and sizes
- Have difficulty comprehending implied meanings in word problems

Description. Students who are visually impaired learn to do mathematics in much the same way as their sighted peers but at a slower pace. To the partially sighted, some print may appear distorted and/or blurred. These students often must carefully study the page to make sense of it, pointing to each number to keep their place on the page. These processes take extra time, slowing the computation rate. Many students also have difficulty interpreting the meanings of word problems.

Causation. Denied the range and quantity of visual stimulation and learning enjoyed by their sighted peers, visually impaired students often lack relative experiences. Additional concrete experiences, including manipulative and tactile examples, must often be provided. Similarly, the experiential background with which to understand and interpret mathematics problems must be carefully built for them. The effort and time involved in deciphering print also seem to interfere with extracting meaning from passages. Shortening word problems may facilitate comprehension as well as prevent students from tiring from the strain. Much the same applies to those who are blind, except that they must rely on hearing, touch, and the mastery of Braille to progress in mathematics.

Implications. Given sufficient auditory support, visually impaired students can steadily improve their mathematics skills. However, their tendency to compute at a slower rate and to take longer to master mathematics skills may prevent them from keeping pace with the typical mathematics curriculum. For the partially sighted, many classroom mathematics tasks—such as copying problems from the board—are particularly difficult to accomplish. These students may fall behind peers both in mathematics progress as well as in demonstrating task mastery.

CORRECTIVE PRINCIPLES Use these guidelines to plan and modify instruction.

1. Seat students where they can best see and hear the teacher.
2. Present auditory cues for each mathematics task.
3. Alternate auditory and visual mathematics tasks.
4. Build both real and vicarious experiences for each mathematics lesson.
5. Provide proper illumination, legible print, and large-print texts as needed.
6. Provide extensive training in technical vocabulary.
7. Use focused listening activities to teach problem-solving skills.

ACADEMICALLY GIFTED LEARNERS

DETECTION Properly identified students may:

- Appear gifted in all areas except mathematics
- Display a high interest or a disdain for mathematics
- Perform drill grudgingly, incompletely, or not at all
- Make numerous careless errors

Description. It is not uncommon for gifted students to read significantly above grade level yet perform mathematics at or just slightly above grade level. This is because certain mathematics concepts and mechanics are difficult to learn independently. Although they typically grasp new concepts quickly and require only limited drill, gifted students' progress in math is often restricted by what they have been directly taught.

Causation. Not all gifted students are gifted mathematicians. Giftedness does not necessarily cut across all curricular areas. Some students have not been taught the concepts and mechanics to advance. Many gifted students prove to themselves that they can perform a task and then lose interest and perform carelessly. Others overlook details in their concern for the larger concepts. Their rapid learning rates often result in disdain for boring drill. Underachievement may result from conformity to school expectations, sociocultural differences, or handicapping conditions that interfere with performance.

Implications. Students who are not gifted in the area of mathematics should be taught in much the same way as regular class students. However, many gifted students can advance rapidly in mathematics skills if they are taught the specific strategies and clues to do so. Often, a quick 15-minute lesson on mechanics can free the students to rapidly progress to more difficult math tasks instead of suffering through additional drill.

CORRECTIVE PRINCIPLES Use these guidelines to plan and modify instruction.

1. Before implementing a special program, establish each student's giftedness in the area of mathematics.
2. Use an individualized mathematics approach.
3. Provide direct instruction in mechanics and concepts for advanced skills.
4. Offer enriched, challenging mathematics experiences.
5. Avoid unnecessary drill, allowing students to advance as they are able.
6. Accept and encourage novel strategies for computing and solving problems.

REFLECTIONS

1. The organization of Part I suggests differences in the mathematics needs of students in regular/remedial classes and those who are eligible for special education. Compare and contrast a problem in each area. Are there distinct differences in the DETECTION behaviors and the CORRECTIVE PRINCIPLES? Why or why not?

2. Similar observable behaviors are cited for several categories of students; scan the DETECTION behaviors to locate commonalities across categories. Which categories cite the most similar behaviors? Why do you think this is so? Follow a similar procedure to compare CORRECTIVE PRINCIPLES across categories.

3. Many of the principles in Part I apply to all students. Justify the selection of the ones presented, adding, deleting, or modifying principles as you deem necessary.

4. Chapter 1 presents an overview of a diagnostic/prescriptive model. From a practical point of view, identify two features of this model that would be difficult to apply in a regular mathematics classroom. How could they be overcome?

5. Students' problems in mathematics tend to assume different proportions according to the student population and the perceptions of individual teachers. Interview a highly skilled regular education teacher to determine his or her perception of the important DETECTION behaviors and CORRECTIVE PRINCIPLES for each categorized problem; then discuss detection and correction of any frequent problems not mentioned in Chapter 1. Follow a similar procedure to interview a veteran special education teacher.

6. Both teaching and learning mathematics are complicated processes. Based on the discussion of special mathematics needs in Chapters 1 through 3 and your experience, for which categories do you think regular classroom mathematics instruction is the most difficult to provide? Why? For which categories is it easiest? Why?

7. Many school systems offer cross-categorical special education services, grouping all mildly handicapped students together for instruction. Reread the DETECTION behaviors and the CORRECTION PRINCIPLES cited for the exceptional classifications; add your own observations. Then debate the value of categorical and noncategorical special education training in mathematics. Consider the advantages and disadvantages of each model to the individual students, to the other students in the same classes, to the teacher, and to the school system.

8. The CORRECTIVE PRINCIPLES are suggested as guides for selecting and modifying mathematics strategies. Select a hypothetical special learner; using the appropriate CORRECTIVE PRINCIPLES as guidelines, plan for that learner a modified basal mathematics lesson. Repeat the process for a second special learner. Compare and contrast the two lessons. For the same content, review the lesson script in the teacher's edition of a basal. How do your lessons differ from the ones suggested for most students?

9. Teachers often know how but don't have time to plan special mathematics lessons for special learners. Volunteer your services to plan one or more special mathematics lessons for a special learner in a nearby classroom. Take the mathematics content of your lesson from the basal or other materials currently in use in that school. Use the diagnostic information available from the school and the appropriate CORRECTIVE PRINCIPLES to guide the design of your lesson.

10. Teachers also have trouble finding enough time to teach special mathematics lessons. Volunteer to actually teach the mathematics lessons you designed.

11. A number of mathematics and special education textbooks address the special mathematics needs of special categories of students. Compare and contrast discussions in these sources with the information in Chapters 2 and 3:

Ashlock, R. B. (1986). *Error patterns in computation* (4th ed.). Columbus, OH: Charles E. Merrill.

Choate, J. S., Bennett, T. Z., Enright, B. E., Miller, L. J., Poteet, J. A., & Rakes, T. A. (1987). *Assessing and programming basic curriculum skills*. Boston: Allyn and Bacon.

Heddens, J. W. (1984). *Today's mathematics*. Chicago: SRA.

Howell, K. W., & Morehead, M. K. (1987). *Curriculum based evaluation for special and remedial education*. Columbus, OH: Charles E. Merrill.

Johnson, S. W. (1979). *Arithmetic and learning disabilities*. Boston: Allyn and Bacon.

Reisman, F. K. (1982). *A guide to the diagnostic teaching of arithmetic*. Columbus, OH: Charles E. Merrill.

Salvia, J., & Ysseldyke, J. E. (1987). *Assessment in special and remedial education* (4th ed.) Boston: Houghton Mifflin.

PART II

PROBLEM-SOLVING NEEDS

Problem solving is considered by many to be the primary function of mathematics education. In the Agenda for the 1980s developed by the National Council for Teachers of Mathematics, problem solving leads the list of areas of greatest need. Students must learn how and when to use the computation and fact skills they develop or those skills will be of no use at all.

Throughout the country, elementary- and intermediate-level students are having significant difficulties with problem solving. Even in situations where students score near or at grade level in mathematics computation, they consistently fall far below the mark in problem solving. In the *Third National Mathematics Assessment: Results, Trends and Issues* (National Assessment of Educational Progress, 1983), the percent of middle-school-age students who were unable to perform basic problem-solving skills ranged from 45% being unable to solve one-step problems using whole numbers to 65% being unable to solve nonroutine problems. Even at the 45% level, nearly half of all students are failing to learn, and that involves only simple problems. This is obviously a serious issue, one that will require a concerted effort for remediation.

How should we approach remedying this situation? Suydam (1982) presented a thorough review of the research on problem solving. Among other things, she suggested that students who are exposed to a wide array of strategies are more likely to use the strategies during problem solving. Thus, a successful problem-solving program should focus on systematically teaching students to build this coherent model or procedure, rather than providing students with a fragmented skills approach, as many of the basal programs do.

The need for systematic teaching is especially profound when dealing with special students, who typically have difficulty organizing the learning process. Special-needs students are not good at taking random pieces of information and integrating that information into their learning system. Although the isolated lessons found in most basals are well developed, they miss the needs of special learners, who cannot fit the lesson into what has come before or what will come after. A fragmented lessons approach tends to confuse rather than enlighten the learner.

A final advantage offered by a systematic approach is applicability. No one strategy is useful in solving all kinds of problems, and different students may apply different strategies to solve the same problem. The specific attack strategy a student chooses is not as important as whether the student uses the strategy in the context of a sound procedural model. Students must become familiar with various kinds of strategies and learn to work in a systematic and organized manner.

Part II focuses on the four student abilities that are basic to problem solving: reading, data organization, operation selection, and answer evaluation. Chapter 4, on reading, examines the ability to read mathematics problems at the independent level. Although this need may be fairly obvious, many students do not read well enough to comprehend mathematics problems; there may be problems with fluency and/or vocabulary. Another aspect of reading comprehension addressed in this chapter is the students' ability to rewrite the main question into their own words, which indicates how well students understand the question being asked. As an extension of rewriting problems, students can create their own problems, thus learning the component parts of a problem and making them aware of what to look for in future problems. (Certainly, an area as significant as reading cannot be covered in one short chapter. The reader is referred to *Reading: Detecting and Correcting Special Needs,* by J. S. Choate and T. A. Rakes [Boston: Allyn and Bacon, 1989].)

Chapter 5 deals with the ability to organize information. Students with special needs often have difficulty sorting out data; in particular, they have problems selecting relevant from extraneous data. Finding facts within problems and then organizing those facts in a coherent manner is difficult for students with sequential learning problems. Likewise, determining the importance of any fact requires the student to make a judgment regarding that fact in a given context. This chapter also suggests how to teach finding and using data from tables and graphs, which has real-life significance, given the current media trend toward graphics.

Chapter 6 focuses on the single greatest mathematics need of all students but especially of learners with special needs: selecting the appropriate operation. Educators have spent a lot of time trying to understand why operation selection is a problem and how this problem can be resolved. The activities presented in this chapter are only meant to begin a most important process; they have been field-tested and thus shown to improve students' understanding of how to pick the correct operation. These activities should be used in the sequence in which they are presented; the amount of time spent on each activity may vary from setting to setting. After students have learned how to pick the correct operation, they must learn how to apply that decision to a working equation; that is the second topic of Chapter 6.

Chapter 7 covers evaluating the correctness of the answer. The main emphasis is on determining how reasonable the answer is. This does not refer to estimating but rather to the manner in which the answer matches the question asked. Students with special needs often lose sight of where they were trying to go during the process of problem solving; consequently, they often end up some place altogether different. The only way for students to be sure that they have followed the correct path is for them to compare their product with the product the problem was seeking. This is not a difficult procedure, but special learners seem to avoid it.

Problem solving is the culmination of mathematics in that all skills are used in a meaningful way. As the central focus of mathematics, problem solving puts meaning to all that we do with numbers. Each new operation is learned in order to somehow facilitate problem solving. Why would one bother to learn to add numbers except to find "how many in all?" The remaining parts of this book deal with skills that assist in problem solving and should be viewed from that perspective.

One last thought on problem solving: As a process, problem solving takes a long time to develop. In fact, it is a process that begins when we first encounter our world and have to make choices between competing alternatives and continues to develop throughout our lives. Hopefully, we become better at the process over time.

Special learners don't learn as easily from life's incidental experiences. For them to learn, experiences need to be somewhat controlled, explained, and repeated over and over until understanding takes place.

With this in mind, consider two points. First of all, you will not help the special learner by saying, "Try and think about it." And second, teachers who work with special students must be prepared to invest a significant amount of time before they are likely to see any progress in students' ability to solve problems.

1. READING FLUENTLY

DETECTION This may be a special skill need of students who:

- Stumble over words while reading problems
- Do not understand what they read
- Ask numerous questions about the meanings of words
- Cannot remember what they have read
- Are reading far below the level of age peers

Description. For the purpose of solving mathematics problems, students must be able to read all material at the independent level. If the material is either at a higher readability level or contains significant vocabulary at a higher level, students will be unable to read the problems and address the primary issue of comprehension. It is even more important to be careful about readability and vocabulary when dealing with students with special needs. If problem-solving instruction is to concentrate on the development of critical thinking skills, the students' attention must not be distracted toward areas that interfere with those skills.

Causation. The causes of reading difficulty are as varied as the number of reading difficulties themselves. A large proportion of students with special needs experience serious problems reading. Again, for a comprehensive treatment of this area, consult alternate references that specifically address the reading needs of special learners.

Implications. Students who have significant difficulties with reading and vocabulary will not be able to utilize the appropriate grade-level materials to learn the necessary problem-solving skills. The teacher will have to scrutinize any material in problem solving that these students might use. One immediate possibility is to present the problems orally or pair special students with peers who can easily read the materials. If students are paired, it is important to monitor the behavior of both team members so that the more successful student does not do all the work.

CORRECTION Modify these strategies according to students' learning needs.

1. *Assessment.* Use an informal reading inventory to more completely evaluate the students' reading difficulties. If this is not possible because of class size or the number of students being served, refer these students to a reading specialist for a complete reading evaluation.

2. *Vocabulary Drill.* Sometimes students can read adequately to work in the appropriate materials but still have trouble with the technical or subtechnical vocabulary used in mathematics problem solving. Develop a set of flashcards of the difficult words for students to learn them as sight words. Have students work with partners in building the sight words. Also have them find the meanings of the words and write each on the reverse side of each card. The students should then use each word in a sentence, which should also be written on the card. The meanings of technical vocabulary can usually be found in the glossary of the basal mathematics book.

3. *Taped Problems.* Recruit volunteers to tape problem-solving lessons that correspond to written materials. These tapes must be carefully coded both on the label and on the recorded message for identification purposes. The recorded message should begin by identifying the name of the lesson, topic, and page number. As the students play these tapes, they can follow along in the written materials, marking on them where necessary. The advantage of the taped lessons is that students can replay them as many times as necessary to get the full meaning of each lesson. Also, by following along in the written materials, students may begin to develop some of the necessary sight vocabulary. Although this may seem like a major effort, remember that you can use volunteers to make the tapes and that the tapes can be used many times.

4. *Book Review.* A necessary step to insuring that your students will be able to use the materials you give them is to evaluate the materials themselves. Carefully read any introductory materials that might be contained in the student books or the teacher's edition to see if readability and vocabulary have been controlled. If these items are not mentioned, they probably have not been addressed; you may need to evaluate each book yourself. You can use any number of readability formulas and/or word lists to complete this procedure, but it is not something that can be accomplished quickly. When you are evaluating materials for possible purchase, you may want to evaluate whether they have controlled reading and use this as a factor in making your decision.

2. UNDERSTANDING THE MAIN QUESTION

DETECTION This may be a special skill need of students who:

- Frequently solve the wrong problem
- Would title a story incorrectly
- Do not get the major message in a story
- Know the facts in a story but can't put them together

Description. After the problem has been read, the first step in problem solving is to locate and understand the main or primary question. The main question is frequently but not always followed by a question mark. It is also frequently but not always at the end of the problem. Students who cannot locate the main question can go no further in solving the problem. The main question in fact determines everything the learner must do throughout the rest of the problem-solving process, including evaluating and organizing data, selecting appropriate operations, and judging any final answer.

Causation. Two issues are central here: locating and then understanding the main question. Locating the main question is usually a mechanical process and one which can be structurally thought out; students who cannot do this have often missed the necessary instruction. Students who do not understand the meaning of the main question may have serious difficulty with reading comprehension in general. The same needs discussed in depth in *Reading: Detecting and Correcting Special Needs* (Choate & Rakes, 1989) apply here. In addition, the students may have conceptual difficulty or confusion in the mathematics domain. That is, they may be fairly competent readers in other areas but have not developed an understanding of the nature of operations in mathematics.

Implications. Students who require assistance to locate the question will need simple instructions on finding the sentence with the question mark at first. Later on, they should be guided to find sentences that pose questions in the form of "if/then." Students who experience more conceptual difficulties will need extensive training in both reading comprehension strategies and basic mathematics principles.

CORRECTION Modify these strategies according to students' learning needs.

1. *Reading Review.* Review the sections on finding the main idea outlined in *Reading: Detecting and Correcting Special Needs* (Choate & Rakes, 1989). Use the strategies that will enhance understanding of the main idea in problem solving.
2. *Question Search.* Have the students practice finding the question when it is clearly stated within a problem and followed by a question mark. Explain to the students that the question is not always followed by that punctuation but that this is frequently the case. If the materials are consumable, have students highlight with a felttip marker first the question mark and then the question itself.
3. *Rewrite the Question.* One way of knowing if the students have a firm grasp of the main idea of a problem is to have them rewrite the question in their own words. Do not have students try to select an operation at this point; this will come later. This activity can be practiced every day, which should increase the students' proficiency. Have them work in teams of 2, and then have the teams compare answers. Students will see that you can use many different combinations of words to say or ask the same thing.
4. *Draw a Picture.* Sometimes students have a partial idea of what is being asked within a problem. When this is the case, it is often useful to have them draw a picture showing the action within the problem. Frequently, this action drawing will help students clarify what the problem is all about. The quality of the artwork is of no concern here at all. Once the picture is drawn, have students find the question the picture illustrates.
5. *Build a Model.* Again, sometimes students have a partial understanding of the problem. In this case, building a model of the problem by moving some set of manipulatives around can help expand that understanding. The objects used are not important to the process. What is important is that the students begin to build a total model, beginning with the pieces that they know, then filling in as the problem becomes clearer. This is an excellent opportunity for 2 students to work together. Since each will probably see the problem from a slightly different perspective, each can add something to the model being built. Encourage students to think aloud (quietly) as they build their models.
6. *Student-Generated Problems.* Having students create their own problems causes them to identify all the main elements in problems. Numerous data banks are available (including almanacs and encyclopedias) as sources of factual information on which to base problems. In addition to these, you can have students bring in facts from newspapers or magazines regarding events they care about. Subjects like sports and hobbies are always good motivators.

3. FINDING DATA IN PROBLEMS

DETECTION This may be a special skill need of students who:

- Cannot identify important details
- Make incomplete lists
- Leave out numbers in sequences
- Do not remember names, places, or things shortly after reading about them

Description. Facts are the details within a problem that bring specificity to the overall plan. Facts are always exact and complete, leaving nothing to speculation. Students who are not successful at sorting out facts will not be able to use the information given to solve the problem. This does not mean that these students cannot think or develop a plan of action. In fact, many people are quite capable of developing plans and dealing in generalities but unable to nail down the details.

Causation. Some students, including many types of special learners, have never learned to be organized. They approach all tasks the same way: rush in, make conjecture, and rush out. These same types of students are the ones who typically have difficulty developing and applying the computational algorithm sequences. These students often get steps and facts out of order; they also frequently leave facts out. They are the students who forget assignments and often do so because they have not developed a system of organization. Lastly, they demonstrate their disorganization in their study habits and most obviously in their notebooks.

Implications. In designing a corrective system for these students, a general program for overall organization with specific strategies for organizing facts is needed. In general, guiding students to organize notebooks and keep records of assignments is helpful. Regarding story problems, teaching students to use organized lists of steps is helpful. Practice in doing task analyses of things students do every day helps them attend to details. Students who have difficulty with identifying salient details will have difficulty with all upper-level mathematics as well as any other subject that requires managing facts and details. Thus, science, especially chemistry, will be difficult for these students.

CORRECTION Modify these strategies according to students' learning needs.

1. *What's in the News?* Get a copy of any local paper and have each student take an article. Have the students underline each fact separately throughout the article. Have them go back and count how many specific facts there were in each article. Make a chart to show the number of facts found in each student's article. Repeat the activity for several days so that students can see facts in chart form as well as learn how many facts are contained in everyday reading material.

2. *Mathematics Problems.* Find some out-of-date mathematics texts that can be consumed. (Usually, these can be found in teachers' workrooms or in central storage rooms.) Have the students find the sections on word problems and go through each problem, underlining each separate fact in each problem. Again use your newspaper chart to record the number of facts in each word problem. You can make the comparison between the difference in number of facts in word problems and in newspaper articles.

3. *List the Facts.* Have the students go back through each word problem and list every fact they found. Have them number each fact and write it on a separate line. This will emphasize the separateness of the facts, showing how each one adds some specific information to a problem, some more useful than others. By listing the facts in an organized fashion, the students will have taken data from a large mass of information, organized it, and evaluated it.

4. *Create Your Own.* Guide students to the endless sources of facts that are available: almanacs, encyclopedias, record books, newspapers, and magazines. Have students work in teams of 2 and select 4 facts on any given topic. Have them take the 4 facts and write a problem using those facts. Have 2 teams exchange problems and find the same 4 facts. Then have the teams use the facts they found in the other's problem to create a different problem. Do this activity each day at the end of class for 5 minutes. After a few days, increase the number of facts to 5 and then 6 or 7. This will drive home the utility of facts to students who just like to look at the "big picture."

5. *Insufficient Data.* This is also a good time to introduce the notion of insufficient data in problems. Use problems in which a key variable is left out. Have the students identify the missing variable and then supply the information. It is important for students to see how the missing variable affects their ability to solve the problem. Have the students create their own problems and then delete a key variable. Have teams exchange problems and try to identify what is missing. If they succeed, have the original team provide the missing variable to help solve the problem.

4. DELETING EXTRANEOUS DATA

DETECTION This may be a special skill need of students who:

- Consistently add in regrouped values more than one time
- Do not compare each fact to the question
- Produce answers that are significantly high or low
- Pull numbers out of context to compute
- Would add unlike terms (apples and oranges)

Description. This task is important because many of the "nonmathematics" problems that students face every day are replete with extraneous data. Problem solving is a process of choosing between and among competing data. The ability to screen out extraneous data is essential to being able to organize information into useful sets for problem solving. Students who cannot recognize extraneous data will fail at problem solving because they are working with the wrong information. And if an answer is somehow derived, students will be unable to effectively compare their answer to the original question with any degree of accuracy. Also, when students do not eliminate extraneous data, their faulty answers might seem reasonable when in fact they are exceedingly high or low.

Causation. Problems with deleting extraneous data can be the result of several factors: 1) The student does not fully understand the main idea (main question) of the problem; 2) the problem is overloaded with factual information; or 3) the student does not attend to the details of the work. The first two causes relate to lowering the difficulty of the work and can therefore be resolved accordingly. Failure to attend to detail is clearly a problem with the learner and can be most effectively corrected by assigning activities that require only fact evaluation.

Implications. Problems with extraneous data are serious because they will affect the total comprehension process within problem solving as well as deter the student from successfully comparing the answer to the question. Students who cannot identify extraneous data will constantly confuse important and unimportant data, not only in problem solving but in all content areas.

CORRECTION Modify these strategies according to students' learning needs.

1. *Concentration.* Select several problems at the students' independent reading level. Write each on an index card. On the reverse side of the card, write 2 facts from the problem and 1 fact that is not found in the problem. Ask students to select the fact that is not found in the original problem.
2. *Fact Hunt.* Select 4–5 problems at the students' independent reading level. Ask students to find all the facts in the problem and then to underline each fact separately. Explain that first you must find the facts before you can decide which are useful in solving a problem; relate this to solving a mystery. Then have students reread the problems and circle only the most important ones.
3. *Reverse Problem Solving.* Using the problems from Activities 1 and 2, have the students select 2–3 facts from each problem and write a totally different problem. This will show students that facts are only meaningful in the context of the problem in which they are found. If a student creates a problem that makes a previously useful fact extraneous, use this situation to amplify the importance of context.
4. *Compare and Contrast.* Select a number of problems at each student's independent reading level. Have the students underline each fact in the problems. Next, have the students circle the main question. Lastly, have them compare each underlined fact to the main question to decide if that fact is related to the question. Guide students to think aloud about the relationships.
5. *Omission.* Write questions on index cards, leaving out a key fact in each one. Leave a blank space where the fact should go as a contextual clue. Write each missing fact on a separate card. Have the students match the cards and explain their choices.
6. *Concentration.* Use several of the cards developed in Activity 5. Turn all the cards face down. Have students take turns, each picking 2 cards a turn; if the cards match, the student gets a point. If they don't match, the cards are turned face down again and the next student gets a turn. The aim is to remember facts and match them to the correct problem cards.

5. LOCATING DATA IN TABLES AND GRAPHS

DETECTION This may be a special skill need of students who:

- Do not use information found in the tables or graphs accompanying text
- Cannot look up times or dates on schedules
- Do not align their numbers in columns

Description. Finding data in tables and graphs is becoming more and more important in today's world because more and more data are being presented in that format. One look at any national magazine or newspaper will confirm this contention. Students who cannot find the data they need clearly will not be able to apply it in problem solving. Tables present data in column format, usually with categories of information on one axis and descriptors on the other axis (columns vs. rows). Graphs present data in a more pictorial way, using a pie-shaped graph to show proportions, a bar graph to show relative values, and a line graph to show linear change. The purpose of each type of data presentation determines which form the data will appear in. If there are a great number of facts about which many different types of comparisons or conclusions can be made, a table will be most useful. If there are fewer details and comparisons and it is more important to look for trends in the data, then a graph is more useful.

Causation. Students who have difficulty finding and utilizing data from tables or graphs are usually not very organized; they often cannot find things and their work is often messy. With regard to problem solving, these students tend to deal with the larger issues of the problem, often neglecting to find or use data to back up their point. Just because these students are not tidy or systematic, we should not conclude that they cannot learn to be. The solution is to help them develop an overall organizational approach. Some special learners, particularly those with visual deficits, find this particularly difficult.

Implications. Students who cannot find data in tables or graphs will be unable to make decisions or draw conclusions from data being presented. This will not only be a problem with mathematics but will affect the sciences and social sciences, where data are so often presented this way. Students need to learn a step-by-step process to find data in various formats.

CORRECTION Modify these strategies according to students' learning needs.

1. *Table List.* Provide students with a step-by-step list to follow in finding data in tables. For example:

    Information groups that run up and down are called *columns*. Information groups that run side to side are called *rows*. Use a ruler or straight edge to help locate information in a given column or row.

    a. Take any table of data.

    b. Have the students find a column.

    c. Have them go down the left-hand column to a row you have identified and look across that row to the column they found in step b.

    d. Have students read that piece of data and label it, using the names of the column and row.

    Remember: When you look up information in a table, move your finger down the first column until you find the name of the person or thing you are looking for. Then move your finger across.

2. *Graph List.* Provide the students with a step-by-step guide to interpret graphs. For example:

    Like a table, a graph shows information in an organized way. Unlike a table, a graph uses a picture to show the information. A graph is helpful for comparing 2 or more facts. This lesson will teach you about a graph:

    a. Find the title of the graph. This will tell you what is being compared in the graph.

    b. Look up and down the left side. This will tell you 1 characteristic being looked at. (For example, it might tell you how many times something took place.)

    c. Look across the bottom of the graph. This will tell you the other characteristic being compared. (For example, it might tell you who did it or what year the event took place.)

    d. Select a graph from a newspaper, and have the students find different facts in it.

    e. Have the students use the data in the graph to make comparisons between 2 or more items in the graph.

3. *My Own.* Have the students work in groups of 2–3 to create tables and graphs of their own using information that comes from school. They could use the number of students that attend, number of absences each day during a week, cafeteria information, library check-out data, and so on.

4. *Place Cards.* Give students who have difficulty tracking information down and across tables and graphs an index card or bookmark to use as a "place card." Have students locate the first descriptor with their finger and the second with the place card.

6. TRANSLATING TERMS INTO ACTION

DETECTION This may be a special skill need of students who:

- Seem to pick an operation at random
- Do not pay attention to the sign
- Have not properly identified the main question
- Do not truly understand the operation and what it accomplishes

Description. When students translate terms to action in operation selection, they are using the language of the problem to determine which operation is needed. To do this, they really must sort the problem into one of several categories and then, in many cases, sort even further as they subdivide the categories into smaller groups. This is the most difficult step in problem solving because it requires the use of higher-level critical thinking skills. Many special-needs students find this sorting process especially difficult and will need extensive guided practice for mastery.

Causation. Students who cannot translate terms into action may have one of a number of problems. First, they may not understand the concept of each operation; that is, they may have learned the steps in division but do not understand that the function of division is to share things equally. Second, students may have a problem understanding the relationship between whole numbers, fractions, and decimals. Third, students may be unable to find the main question. And lastly, they may not have the necessary vocabulary. As a critical thinking skill, this is a particularly difficult task for students with limited mental abilities.

Implications. Students who cannot select the correct operation simply cannot solve the problems. Calculators will not help them, nor will computers. Students must master the critical-thinking ability to sort problems into categories and then subsort into additional categories as needed.

CORRECTION Modify these strategies according to students' learning needs.

1. *Check Computation Skills.* Review Chapters 9 through 12 to be sure the students can perform the needed operations. Use a criterion-referenced test to be sure the students have mastered necessary computation skills.
2. *Find the Question.* Copy or have students copy 15–20 questions out of their basal mathematics text. Have them write the main question on the back of the card. Then have them shuffle the cards, and, working in groups of 2, have them practice finding the question. Groups can exchange decks of cards so that students do not practice with the same deck twice.
3. *Vocabulary.* Explain to the students the words *total, compare, distribute,* and *difference.* Make up poster cards with these words, and place them on 4 different tables or desks in your room. Read one of the problems from the cards the students used in Activity 2. Select a student and have him or her decide which of the tables it should go on. Select about 8–10 examples of each type and go through this sorting process. Make sure to mix up the cards so there are not several of one kind together.
4. *Team Sort.* Have the teams of 2 students go through the rest of their decks of cards, sorting the problems into the proper categories. Have the teams exchange decks over and over so that each team could end up with as many as 15 decks of cards. This practice is critical to developing the thinking skills needed to select the correct operation. Take all the time you need to develop these skills, as they are the most essential of all the skills in problem solving.
5. *Subsort.* This activity is difficult and will have to be covered slowly with systematic review built in. Take 1 category from Activity 4 above. Take 2 problems, 1 for addition and 1 for multiplication. Discuss how they both find totals. Next, list the facts for each problem, crossing out any extraneous facts. Look at each fact and label it as an "addend" or a "factor." If all needed facts are addends, put that card in the "add" pile. If one of the facts is a factor, put that card in the "multiply" pile. Only use 1-step problems at this time. Next, do this for comparative problems and distributive problems. Think aloud as you model each step. After you model the procedure several times, have students follow your model, thinking aloud as they work.

7. WRITING EQUATIONS

DETECTION This may be a special skill need of students who:

- Write down one of the facts as the answer
- Use a calculator and push an operation at random
- Write incomplete or grammatically incorrect sentences
- Frequently act without a plan

Description. Once students have learned to select the correct operation, they must learn to put that operation and the corresponding facts into a statement called an equation. Some argue that an equation is really an open-ended sentence, but students must learn to write the equation as a complete statement; the concept is much like that of writing a sentence with a subject and predicate. An equation must have two sides. On one side of the equals sign, the action is demonstrated, showing the relationship between facts. On the other side of the equals, a statement of the consequence is formed, worded in the same way the answer would be but without the numerical component. The consequence should be compared carefully to the main question at this point, as this would affect the ultimate outcome; the statement of consequence should in effect answer the main question of the problem. Thus, the relationship set up on one side of the equals sign results in the consequence on the other side, and the consequence answers the question that was asked in the problem.

Causation. Frequently, students who have the greatest problem with constructing equations do not grasp the full concept of cause and effect. They cannot follow the logic of an "if/then" relationship. While this logic can be explained and practiced, there is also a developmental element involved: Many students who have moderate or greater intellectual limitations will take a long time to understand cause/effect relationships. They can learn, however, and part of that learning will be a function of how the concept is presented to them.

Implications. Students who cannot generate an equation will have significant problems trying to put facts into proper order before computing. These students will also have even greater difficulty as they progress to higher-level courses in math and other subjects that require the organizational skills represented in equation generation. Developing these equation-writing skills must be approached in a step-by-step manner.

CORRECTION Modify these strategies according to students' learning needs.

1. *Use Equations.* Select sample equations already available in the math texts. Explain each equation and its parts and their functions. Read the problem as the relationship of fact A to fact B, resulting in the answer. Next select 4–5 of the problems from the cards generated for Special Need 6. Write an equation for each card, again reading the equation as a function between fact A and fact B resulting in an answer. Show the students how this is similar to equations they have been using already. Focus on 1 operation and repeat the process several times.

2. *Teamwork.* Have teams of 2 students take a card from the total pile. Have them review the problem, find the question, see why it was in the total pile, and list the relevant facts. Provide a blank equation format (____ + ____ = ____). Have the students fill in the appropriate parts. Practice this for several days.

3. *Difference Equations.* Next have the students select a few sample questions from the difference pile. Provide them with an appropriate blank format (____ — ____ = ____). Have them review the problem, find the question, see why it is a difference problem, list the relevant facts, and fill in the equation. Practice this for a few days.

4. *Mix and Match.* Provide the blank formats of Activities 2 and 3 above. Mix problems from the 2 piles. Have the teams examine each problem and sort it into the appropriate operation and then go through the steps listed in Activities 2 and 3.

5. *Move On.* Add the category of multiplication and the appropriate blank format for the equation. Have the teams go through the steps outlined and fill in the parts. Next provide blank formats for 3 equations and mix the new category in with the previous 2. Have the students go through the same process as Activity 4.

Note: It is very important to provide this kind of practice in isolation and then build systematically to what has already been learned. Most available programs do not do this and in fact leave this up to the student, who is totally unprepared for this requirement. Presenting skills one at a time is essential for the development of learners who have special needs. However, equally important is the need to assist these learners in synthesizing these skills.

8. COMPARING ANSWER TO QUESTION

DETECTION This may be a special skill need of students who:

- Produce answers that do not match the questions
- Frequently give answers in class that do not relate to the questions
- Often lose their place in the lesson
- Do not pay attention to detail
- Have a short attention span

Description. Frequently, low scores on achievement and other tests are related to the students' lack of attention to detail and poor proofreading skills. It is necessary in problem solving to be sure that the question you have answered is the question that was asked in the first place. Sometimes, students go off on tangents and return at the wrong point; they then move on through the problem and end up having solved a different problem. One way to check this is to stop at some midpoint and compare the answer to the original question. Again, the emphasis is on how all the steps in problem solving build on one another; thus, the very first step of understanding the problem is essential to the end of the process.

Causation. Students often feel they need to rush through a large number of problems when quantity rather than quality is stressed in the class. And some students, particularly special learners, compute very slowly and spend a disproportionate amount of their problem-solving time with this mechanical step of answer/question comparison. In addition, some students have simply learned over time to be careless.

Implications. Students must learn to check their work; certainly, when they leave the arena of the classroom, no one will do it for them. Teachers should also stress that it is not the sheer number of problems completed that counts but rather the number completed correctly over the number attempted. After students have learned the correct process, they will have plenty of time to increase their pace of production. One way to relieve the burden of computation for those students who compute slowly or incorrectly is to provide them with calculators.

CORRECTION Modify these strategies according to students' learning needs.

1. *Check Computation.* Again, this is an appropriate place to thoroughly check each student's ability to perform all necessary operations.
2. *Calculator Time.* If the students are not familiar with calculators, introduce them. Carefully demonstrate how to use a calculator for each separate operation. Provide plenty of practice, as this powerful tool can be very helpful if used correctly and very dangerous if used incorrectly.
3. *Match Time.* Use the problem cards developed in Chapter 6. Create 2–3 answers for each problem. Only 1 of the answers should match the question in the problem. Have the teams brainstorm and try to pick the correct answer without calculating. To do this, they will study the relationship of each possible answer to the question. Next have them check themselves by solving the problem.
4. *Estimation.* Read aloud problems and have students race to estimate the answer. Have them explain why they think their answer is fairly accurate. Then, as in Activity 3, give them 2–3 choices, none of which is precisely correct, to compare to the estimations. Then have students actually solve the problems to confirm estimations.
5. *Extended Practice.* As this is the last step in problem solving, have the students practice solving problems and checking their answers. They should go through the process of carefully studying the problem, finding the question, selecting useful facts, determining an operation, generating the equation, calculating the answer, and checking that answer. This process should be repeated over and over until it becomes second nature to the students.

Note: Since a significant amount of practice is required, base the problems on topics of interest to the students. Ask your students what they enjoy reading about.

REFLECTIONS

1. Try some of the strategies suggested in Part II to create problems for a problem-solving bank. If you have your own class, have your students help by contributing problems to the bank. They should develop problems around areas of key interest to them. If they are hesitant, suggest sports or hobbies to get them started; then try to get students to move on to other areas, such as current events. Use the bank for different activities in class.

2. Related observable behaviors and skills are cited for the steps of problem solving; scan each discussion to locate these relationships. Which steps cite the most similar behaviors and skills? Why do you think this is so? Follow a similar procedure to compare CORRECTION strategies across the problem solving steps.

3. Many of the CORRECTION strategies apply to all students. Justify the selection of the ones presented, adding, deleting, or modifying strategies where you deem necessary.

4. Problems in mathematics tend to assume different proportions according to the student population and the perceptions of individual teachers. Interview a highly skilled regular education teacher to determine his or her perception of the important DETECTION behaviors and CORRECTION strategies for each categorized problem; then discuss detection and correction of any frequent problems that are not mentioned in Chapters 4 to 7. Follow a similar procedure to interview a veteran special education teacher.

5. Both teaching and learning mathematics are complicated processes. Based on the discussions of special mathematics needs in the previous chapters and your experience, for which difficulties in problem solving do you think regular classroom mathematics instruction is the most difficult to provide? Why? For which is it easiest? Why?

6. The CORRECTIVE PRINCIPLES in Part I are suggested as guides for selecting and modifying the mathematics strategies given in Part II. Select a hypothetical special learner; using the appropriate CORRECTIVE PRINCIPLES as guidelines, plan for that learner a modified problem-solving lesson. Repeat the process for a second special learner. Compare and contrast the two lessons. For the same content, review the lesson script in the teacher's edition of a basal. How do your lessons differ from the ones suggested for most students?

7. Teachers often know how but don't have time to plan special mathematics lessons for special learners. Volunteer your services to plan one or more special problem-solving lessons for a special learner in a nearby classroom. Take the mathematics content of your lesson from the basal or other materials currently in use in that school. Use the diagnostic information available from the school and the CORRECTIVE PRINCIPLES to guide the design of your lesson.

8. Clip out current event articles from your local newspaper. Group the clippings around common themes or areas of interest. Have students use the clippings to find as many facts as possible. Then have them create many problems using those facts.
9. A number of mathematics and special education textbooks address the mathematics needs of special learners. Compare and contrast discussions in these sources with the information in Chapters 3 through 7:

Choate, J. S., & Rakes, T. A. (1989). *Reading: Detecting and correcting special needs.* Boston: Allyn and Bacon.

Choate, J. S., Bennett, T. Z., Enright, B. E., Miller, L. J., Poteet, J. A., & Rakes, T. A. (1987). *Assessing and programming basic curriculum skills.* Boston: Allyn and Bacon.

Enright, B. E. (1987). *SOLVE: Action problem solving* (Books I–III). N. Billerica, MA: Curriculum Associates.

Heddens, J. W. (1984). *Today's mathematics.* Chicago: SRA.

Howell, K. W., & Morehead, M. K. (1987). *Curriculum based evaluation for special and remedial education.* Columbus, OH: Charles E. Merrill.

Lacy, L., Marrapodi, M., Wantuck, L., Wickenden, B., Witten, T., & Zimmermann, M. (Consultants). (1983). *Problem solving in math* (Books A–D). Cleveland: Modern Curriculum Press.

National Assessment of Educational Progress. (1983). *The third national mathematics assessment: Results, trends and issues.* Princeton, NJ: Educational Testing Service.

Panchyshyn, R., & Monroe, E. E. (1986). *Developing key concepts for solving word problems.* Baldwin, NY: Barnell Loft, Ltd.

Suydam, M. N. (1982). Update on research on problem solving: Implications for classroom instruction. *Arithmetic Teacher,* 29 (February 1982), 56–62.

10. *Recognizing more or less:* This primary notion sets the stage for introducing addition and subtraction and also prepares learners in problem solving to ask, "What operation would I need to get more/less?" Although special learners can often perform this skill when the differences are gross, they frequently have difficulty when the differences are more subtle.

11. *Counting objects to 10:* This skill emphasizes the one-to-one correspondence relationship. Many young children count faster than they relate; this becomes a special need when it persists into first grade or beyond. To build this skill, special learners need to have the objects presented in a straight line; when objects are scattered, it is difficult for special learners to remember if a specific item has already been counted.

12. *Joining sets to 10:* Obviously, this cannot be done if the learner cannot count objects to 10 (SN 10). Likewise, students with special needs often do not understand the joining of groups. They frequently see each group as an isolated entity and must develop the relationship.

13. *Demonstrating number concepts to 10:* Students need to be able to understand numbers and number value. As one child said, "1 + 1 = 2. What's a two?" Students must have a sense of what a *two* is. Moreover, they should be able to move from this to summatizing groups of objects.

14. & 15. *Recognizing and writing numbers to 100:* To move into more operational mathematics, students need to recognize and write numbers; eventually, they must do this all the way to 100.

16. *Summatizing to 10:* Students need to be able to recognize a set of objects and assign a quantity to that set. This ability is essential in bridging the gap between counting and number-fact memory.

17. *Attending to sign:* Likewise, students need to pay attention to detail, in particular, the computation sign in the problem.

18. *Applying directionality:* The ability to move in various directions with purpose is a fairly mature skill. Students often learn to add from left to right because that is the direction in which reading takes place. This frequently goes unnoticed because it will not affect the answer until regrouping is introduced much later in the curriculum.

The development of these skills takes a considerable amount of time. As with any concept, the learner must have a sufficient amount of time to experiment with all the ideas in this chapter; nothing will come in a flash with one experience. Frequently, teachers engaged in remedial/corrective education want students to learn more quickly than they can. Once students have mastered the basics, they may learn the mechanics of problem solving more quickly. But at this level of readiness, time is well spent practicing and experimenting.

These readiness skills might seem immature to some teachers, and the urge to replace them with survival skills, especially for secondary students, might be strong. However, even for older students, it is highly doubtful that they will develop many survival skills without having mastered the prerequisite readiness skills. While the teacher may choose to teach calculating with a calculator, if the student does not understand the basic concepts of addition, the calculator will be useless.

In sum, students must develop readiness skills before moving on. If they do not, they will surely experience problems in mathematics and other subject areas.

The real problem for these youngsters will not be that they will not learn. The real problem will be that because they are unable to learn correctly, they will develop misconceptions about mathematics and fall into error patterns that will follow them for years. Those students who are not ready should wait, rather than learn incorrectly.

9. IDENTIFYING SETS OF OBJECTS

DETECTION This may be a special skill need of students who:

- Treat all objects as individual items
- Do not pull a number of objects together to match a given number
- Do not relate a number to a total number of objects
- Do not see the common characteristics of objects in a set

Description. Basic set theory should be learned early in the developmental process of learning basic mathematics. An integral part of that theory involves the recognition of sets, their characteristics, their numerical value, and factors that would separate them from other things. The concept of a set is probably one of the easiest students will deal with. Almost all children have dealt with a set of something before entering school: a set of trains, a deck of cards, or any group of toys that had something in common. Some learners, however, have not learned to group things based on any set of principles. The children who naturally organize things into neat groups will have far less difficulty learning about sets than those who pile everything together. The message behind the concept is that there are rules and unifying principles around which the whole logic of math is going to grow.

Causation. The most likely cause of difficulty with sets is lack of exposure. Many children come to school with little or no experience with the basic mathematics of the world around them. They may spend a tremendous amount of time on nursery rhymes but very little on sorting things or discussing the relationships of objects from a numeric point of view. Developmentally delayed students may also experience difficulty.

Implications. The most important implication of this difficulty is that these students will obviously have serious trouble going on in mathematics. If students do not understand sets, adding and subtracting will make no sense. Fortunately, the instructional implication is clear and positive: Students who have difficulty identifying sets need to be brought back to sorting objects according to common traits. Later, in problem solving, these students will have to apply this same type of skill, categorizing problems as the central step to successful problem solving.

CORRECTION Modify these strategies according to students' learning needs.

1. *Common Trait.* Put small sets of familiar objects with very obvious common traits on a table. Working with small groups of students, have them tell you about the group and give you characteristics of the objects. Many of these descriptions will vary, but list as many as are accurate. Review these characteristics with the class. Help them discover which 1–2 characteristics are the same. Explain to the group that the reason the objects are together is because they have this 1 thing in common.

2. *Include/Exclude.* Leave the small group of objects on the table; put a larger group of objects on the side. Take the new objects 1 at a time and show them to the group. Discuss characteristics of the new object. Have the group decide if it fits with the set on the table. Why or why not? Repeat this for each item.

3. *Set Mates.* Place 2–3 unlike objects on the table. Guide students to explain the differences. Then name each item by a general feature, such as color or shape. Place additional items on the table that share a general feature of one of the named objects. Have the students discuss and then place each new item on the table beside the named object with a similar feature to form sets.

4. *Sort.* Put a number of objects on the table that should fit into 2 different groups. Have the students sort the items into the 2 sets and identify the unifying characteristic of each group. Extend this to items that would sort into 3–4 sets.

5. *Extended Practice.* Repeat Activities 1, 2, and 3 many times until students have a clear understanding of sets and the notion of unifying characteristics. These concepts are absolutely essential to later mathematics.

10. RECOGNIZING MORE AND LESS

DETECTION This may be a special skill need of students who:

- Do not relate the size of a group to value
- Do not put things in proper order
- Do not sequence events correctly
- Do not recognize the signs for less than and greater than
- Do not use space with any sense of proportion

Description.   Recognizing more and less is a truly basic concept. Children often address this as bigger or smaller. For instance, at an early age, children see size as the determining factor in age; they may be confused as to why their grandparents are older if their parents are taller, since taller should equal older. As children grow, they begin to understand that other characteristics determine age, rather than just height. Similarly, as children enter school and deal with money, they start to see that size is not the only factor; later, value becomes the deciding factor. The central process involved here is comparison making. This can build on the skills of identifying sets, Special Need 9.

Causation.   The most likely cause of difficulty in this area, as well as in identifying sets, is lack of exposure. Many children come to school with little or no experience with basic, real-life mathematics. Developmentally delayed students may have particular difficulty.

Implication.   As in the case of set identification, the most serious implication of this difficulty is that students will obviously experience serious trouble going on in mathematics. Again, the instructional implication is clear and positive. Over time, children will learn to make comparisons between objects, moving from comparing based on size to comparing based on other values.

CORRECTION Modify these strategies according to students' learning needs.

1. *Size Difference.* Select 2 students who are very different heights. Have them stand next to each other. Ask the class to decide who is bigger. Repeat this many times, but each time, use a different student such that the students' heights get closer and closer. This will help the students make finer and finer comparisons.

2. *Pile High.* Pile 2 sets of dominoes next to each other on a table. Have the class pick the set that is bigger. Then, count the number of dominoes in each pile and record the comparison on the board, using less than and greater than signs between the 2 values.

3. *Edible Differences.* Bring 2 sizes of cookies or candy to class. (Children instantly recognize the differences in sweet treats!) Use these in either of 2 ways: a) to illustrate direct size comparisons or b) as a more advanced concept, to illustrate the interactions of size and quantity. Reward students who answer correctly by permitting them to eat the identified size and/or number of treats.

4. *Change the Piles.* Put 2 piles of different-sized objects next to each other on the table. Rather than ask the students which pile is bigger, ask which has a greater number of objects. At first, students may pick the taller pile but go back and count, as in Activity 2. Record these numbers on the board with the appropriate sign (< or >), and discuss the idea of number of objects as a deciding factor.

5. *Extended Practice.* Repeat the above activities many times. Use a process called spiraling, where you go through each activity and then go back through all the activities. In other words, rather than wait for mastery at one level, begin the process, move up, and then move back to the beginning. Many young learners need time to incorporate these notions. This process of spiraling, especially at these readiness levels, will help make this basic learning permanent and central to the students' future progress.

11. COUNTING OBJECTS TO 10

DETECTION This may be a special skill need of students who:

- Count very fast and miss objects
- Count slowly and leave out numbers
- Do not address the objects as a group at all
- Look at the group and make up a number
- Look at the group and say "a lot" or something similar

Description. Children learn to apply their rote counting to objects at an early age. As they move from forming groups, they want to know how many are in a group. This process of counting objects is the way that children establish one-to-one correspondence. That notion is essential to later moving up and down a numberline, which sets the stage for addition and subtraction. Obviously, time spent counting objects is well spent. Very young children who first approach this task speed rote count and do not make the one-to-one correspondence. They do not yet really understand that they can determine how many objects are in a group.

Causation. The most likely cause of difficulty in counting to 10 is lack of exposure. Children with little or no experience with the basic mathematics nature of the world most likely do not understand the relationship of objects from a numeric point of view. And those who are mentally and physically immature may not have absorbed the mathematics experiences available to them.

Implications. Students who cannot count objects will obviously experience serious trouble progressing in mathematics. Corrective instruction must encourage counting by actually touching each object, thus slowing down the mindless speed counting. As reflected in the CORRECTION activities, organize the counting into straight lines or simple groups at first, mixing things up later.

CORRECTION Modify these strategies according to students' learning needs.

1. *Straight Line.* The first step in counting objects is to line up 3–4 objects in a straight line. Have the students count the items, touching each item as they go. This touching will stop the speed rote counting mentioned earlier. Increase the number of items as students progress.
2. *Mix-Up.* As students become proficient in using the straight line, move some of the objects around. Do not mix them all up at first, but rather make the line a little wavy and then more and more scattered. The idea is to finally end up with a group of objects that students can count in any arrangement.
3. *Make a Line.* Give the students a group of objects and have them put the objects into a straight line. Have students count the objects before and after rearrangement. This helps the students learn order and organization.
4. *Touch and Tell.* Have students touch as they "tell" or count aloud objects in the classroom, such as desks in a row, books on a desk, windows, or other students. As an extension of Activity 3, you can even have them straighten the room or themselves as they count and Make a Line.
5. *Compare.* Make 2 sets of objects. Have the students count the objects in each set and make a comparison. They should be able to use the skills they developed in Special Need 10 to make the comparisons and explain or place the sign appropriately.
6. *Extended Practice.* Repeat the above activities many times. Use the spiraling process, where you go through each activity and then go back through all the activities. Rather than wait for mastery at one level, begin the process, move up, and then move back to the beginning. Many young learners need time to incorporate these notions; bringing them up again and again after time for thinking is very helpful.

12. JOINING SETS TO 10

DETECTION This may be a special skill need of students who:

- Treat all groups as individual sets
- Do not compare groups correctly
- Do not identify common characteristics of groups
- Cannot break groups down into subgroups
- Cannot remove a subset from a set and have anything left

Description. Physically joining sets is the behavior that must come before meaningful addition and or subtraction. Students start off by counting one set, then the other, and lastly the sets together. This is a simple but very important behavior that is all too often overlooked or quickly run through in the regular mathematics program. Many teachers assume that children come to school knowing this subskill, but like other readiness skills, many children are not exposed to joining sets prior to formal education. Joining sets can also lead to taking sets apart, a practice that is a forerunner to subtraction.

Causation. The most likely cause of difficulty in this area is lack of exposure. Children with deficient basic mathematics experience will be unable to join and unjoin sets, since they do not understand the numeric relationships of objects. Students who are developmentally delayed or who do not have a firm grasp of Special Needs 9–11 will have particular difficulty with this skill.

Implications. Because joining sets has direct implications for adding and subtracting, these students will obviously have serious trouble going on in mathematics. More advanced addition and beginning subtraction make no sense when set joining has not been mastered. To join sets, students must often slow down the counting process so that they can touch each object. Later, they may be able to count solely with their eyes. One very critical step in this process is to have them begin to count objects starting at a given number.

CORRECTION Modify these strategies according to students' learning needs.

1. *Join 'Em Up.* Using concrete material like sticks, buttons, or dominoes, make 2 groups with 2–4 objects in a group. Have the students count the number of objects in group 1 and then count the number of objects in group 2. Have them put all the objects together in a third container and count them again. Record the values for each counting on the board. Move to larger numbers per set and then larger numbers of sets.

2. *Take 'Em Apart.* Make a set of 4 objects on the table. Separate the set into 2 sets, with 1 object in one set and 3 objects in the other. Count the objects and show that there are still 4 objects. Nothing was lost as the groups were separated. Change the total number of objects and the number of objects that go into each subset.

3. *Numberless Line.* Draw a numberline without numbers on it. Starting at the left, count 5 places and mark it. Count another 3 places and mark it. Bracket all the counted spaces and go back and count the numberless points again. This also can work in the reverse way. Start with a number of spaces bracketed and break that down into smaller sets. This will help move students from the concrete into the semiconcrete and later onto paper.

4. *Start With.* One thing that is difficult for students is to start counting at a specific point. For example, students have 2 sets. They count set 1 and get 5. They count set 2 and get 3. To get the total, they go back to recount all the items. Students must learn to start with the total of set 1 when they begin to count set 2. They must say to themselves, "Hold 5 on my brain, now count 6, 7, 8." When students with special needs are adding, they count the first set out on their fingers, then they count the second set on additional fingers, and finally they go back and count all the fingers. The first step needs to be replaced by holding that number in their minds.

5. *Extended Practice.* Repeat the above activities many times, varying the objects to be counted. Practice spiraling, where you go through each of the activities and then go back through all the activities. Many young learners need time to incorporate these notions. Bringing them up again and again after time for thinking is very helpful. When learning readiness skills, spiraling will help establish this basic learning.

13. DEMONSTRATING NUMBER CONCEPTS TO 10

DETECTION This may be a special skill need of students who:

- Give the wrong number of objects when making a request
- Cannot join sets without using concrete objects
- Do not count objects accurately
- Add numbers together by writing both parts as the answer
  (*Example:* 2 + 2 = 22)

Description. Demonstrating number concepts means understanding that five objects equal the number five and that the number five equals five things. This demonstration is the expressive form of counting objects. In counting, we are identifying what is given to us, whereas in demonstrating number concepts, we are expressing in concrete form a number requested of us. Students who become good at this are able to show numbers of objects almost spontaneously. In response to "show five fingers," they immediately hold up five fingers, whereas the beginner counts out the five fingers.

Causation. As with other readiness skills, the most likely cause of trouble in this area is lack of exposure. Inadequate mastery of Special Needs 9–12 and developmental delays may also create problems with demonstrating number concepts.

Implications. Students with this difficulty will obviously experience serious trouble progressing in mathematics. Adding and subtracting will make no sense when students cannot visualize the number concepts. For remedial instruction, students can review the activities practiced here and then build on them to begin the expressive process.

CORRECTION Modify these strategies according to students' learning needs.

1. *Straight Line.* The first step in counting objects is to line them up in a straight line. Begin with 3–4 objects. Have the students count the items, touching each item as they go. This touching will stop the speed rote counting mentioned earlier. Increase the number of items after a few tries.

2. *Numberless Line.* Draw a numberline with no numbers on it. Starting at the left, count 5 places and mark it. Count another 3 places and mark it. Bracket all the counted spaces and go back and count the numberless points again. This also can work in the reverse way, starting with a number of spaces bracketed and breaking that down into smaller sets. This will help move students from the concrete into the semiconcrete and later onto paper.

3. *Start With.* One thing students find difficult is to start counting at a certain point. For example, students have 2 sets. They count set 1 and get 5. They count set 2 and get 3. Then they go back to recount all the items and get a total. To count set 2, they need to start with the total of set 1: "Hold 5 on my brain, now count 6, 7, 8." You will notice that when students with special needs are adding, they count the first set on their fingers, then they count the second set on additional fingers, and then they go back and count all the fingers. The first step needs to be replaced by holding the number from set 1.

4. *Extended Practice.* Repeat the above activities many times. Rather than wait for mastery at one level, practice spiraling, where you begin the process, move up, and then move back to the beginning. Spiraling will help make this basic learning permanent and central to the students' future progress.

14. RECOGNIZING NUMBERS TO 100

DETECTION This may be a special skill need of students who:

- Do not read numbers correctly
- Make numerous mistakes in oral reading of numbers
- Record unreasonable answers in simple computation
- Demonstrate serious and consistent mispronunciations of numbers

Description. Recognizing the numbers 1 to 100 is a skill needed prior to any meaning-ful computation practice or instruction. Recognizing numbers is exactly what it seems: A student looks at the number and says its name. This is very similar to sight vocabulary, where students look at a word and instantly pronounce it. As with sight vocabulary, it is important for this set of behaviors to be automatic. To check this, the numbers should be pre-sented out of order so students do not simply recite. This is not the same as number computation, where students can show the appropriate num-ber of objects. With numbers over 10, this will be a skill that comes later in the curriculum. Likewise, students do not, at this time, have to be able to break the number out into expanded notation. This also will come later in the curriculum.

Causation. Children with limited exposure to the basic mathematics nature of the world will probably have problems recognizing numbers. Many children have not studied numbers per se and few have practiced with numbers to 100. Another major cause of failure at this point is that students may be moved too quickly out of practice on number recognition. This is not to say that the concepts are not critical; rather, additional emphasis needs to be put on factual information.

Implications. Problems with number recognition will impede future progress in math-ematics. Students who cannot recognize numbers will be unable to add and subtract, especially when working with larger quantities. As with any sight- or visually–oriented skill, the immediate instructional focus must be a quick and sight-oriented drill program. Look for an immediate response to a visual stimulus. Therefore, do not encourage long delays in answering because other behaviors might be triggered.

CORRECTION Modify these strategies according to students' learning needs.

1. *Flashcards to 10.* Develop a deck of flashcards with the numbers 1–10. On the back of each card, draw the correct number of items. After the student says the number, he or she can flip the card and self-correct by counting the objects. This activity can be used to teach number comprehension and/or to reinforce number recognition.

2. *Flashcards to 20.* Create a second deck of cards with the numbers 1–20. Again draw the number of objects on the reverse side. So that students will not be overwhelmed with too much information at the beginning, this second deck should be presented only after mastery of the numbers 1–10.

3. *Flashcards to 100.* Develop a set of flashcards to 100. Do not draw the number of objects drawn on the reverse side, as this will distract the student. Because no self-checking device is included, an adult or able peer will need to flash these cards for students to practice.

4. *Paired Drill.* Pair your students into teams. Have them take turns flashing through card decks at the beginning of class as a warm-up activity.

5. *Flash War.* Have the teams challenge each other to quick flashcard wars at the end of class. Keep track of each team's positive answers. As the number of positive answers increases in the same time frame (say, 2 minutes), you will have continuous feedback on learning. Chart the progress of each team.

6. *Extended Practice.* Repeat the above activities many times. Review using the spiraling process, where you go through each activity and then go back through all the activities. Rather than wait for mastery at one level, begin the process, move up, and then move back to the beginning. This will give young learners the time needed to incorporate these notions as part of their basic learning.

15. WRITING NUMBERS TO 100

DETECTION This may be a special skill need of students who:

- Do not write simple numbers below 20
- Misread numbers
- Mix up letters and words when writing
- Can count but not write the corresponding number
- Have difficulty with most writing tasks

Description.   Writing the numbers from 0 to 100 is a critical expressive skill. It requires recognition of the number/name relationship, number comprehension, and the graphics of number formation. Number formation is primarily a mechanical skill. However, the understanding behind the writing is important. Few programs build both understanding and formation together in a rigorous way. Students with special needs must learn the relationship between the formation and understanding from the beginning and then continuously build the skills. Therefore, as they are learning the number *nine*, for example, they need to be reviewing the numbers *one* through *eight.* If the students learn the numbers one through ten early in their program, they will have set the stage for the expansion of this set of numbers to 100.

Causation.   The most likely cause of difficulty in this area is lack of exposure to both mathematics and paper/pencil activities. Developmental delays, particularly in the visual motor area, can also seriously interfere with mastery of this skill.

Implications.   A student cannot move on to computation skills without first mastering number formation. One general source of remedial materials is a program entitled *My Number Book.* This is particularly thorough and interesting for young students as a preventive as well as a corrective vehicle. In addition, students must work on writing their numbers in and out of order. Although it might seem boring to write the numbers to 10 or 20 or so on, the students working on readiness skills seem both to enjoy and need the practice. It can be an activity later put into the form of a race.

CORRECTION Modify these strategies according to students' learning needs.

1. *Trace It.* Make a master sheet with the number 1 repeated on it several times in fairly large print. Review the correct movement of the pencil in writing the number 1, from top to bottom. It may even be necessary to draw an arrow showing direction on the paper. Have students trace copies of this for a couple of days. Next reduce the size of the number to a more appropriate size and add more practices per page. Move from a solid tracing line, to a dotted line, to separated dots, to a starting dot. Go on to number 2, but build in review for number 1. Move on to 3 with review built in for numbers 1 and 2. Do this for all numbers to 10. Have students say the names of the numbers as they write them.

2. *Go Up.* Apply the same set of steps used in Activity 1 to the numbers 1–20. Make sure to review the numbers preceding the number being studied at all times.

3. *Concept Sure.* Be sure that the students know what the numbers they are learning stand for. Work on matching the written number to a corresponding set of objects. Use practice pages that just have a number of objects on them and have the students write and say the corresponding numeral form.

4. *20 Up.* Introduce the tens numbers to 100, one at a time: 20, 30, 40, 50, 60, 70, 80, 90. Demonstrate to the students how to build a higher number: "After 19 comes 20. What comes after 20? Is 20 plus 1, or twenty-one, written 21?" What is interesting is that when students pass 29, instead of 30, they often say "twenty-ten." Take that opportunity to show that the number "twenty-ten" is 30 and "thirty-ten" is 40. Usually, if this is made clear for these two examples, the rest will fall in place. Follow the same procedures used in Activities 1 and 2 to teach the number formations.

5. *Extended Practice.* Repeat the above activities many times. Review with the spiraling process, where you go through each activity and then go back through all the activities. Such review will help make this basic learning permanent and central to the students' future progress. Also, you might use activities from *My Number Book*, by L. Krampe. (National School Services, 632 S. Wheeling, Wheeling, IL 60090).

16. SUMMATIZING TO 10

DETECTION This may be a special skill need of students who:

- Cannot show a set number of fingers quickly
- Do not see a set and know its value quickly
- Must count objects to tell the value of a set

Description. Summatizing is a skill by which learners are able to see a set of objects and call their value instantly. It is not unlike sight vocabulary or number recognition in that it is a very visual skill. However, the major difference is that when summatizing, students have actual items, not numbers or words, in front of them. Summatizing is a critical prerequisite skill to number-fact acquisition. It also offers support for starting to count at a preset place on a numberline. Summatizing can be a fun, quick drill lasting no longer than 30 seconds at the beginning of math class; the teacher holds up a number of fingers for a second and the class calls the number name. Even given all of these applications, summatizing is one of the least practiced skills and activities in elementary-level mathematics classes.

Causation. Limited mathematics experience and visual deficits are likely causes of difficulty in this area. Inadequate mastery of prerequisite skills, Special Needs 9–15, and developmental delays can also interfere with performance of this skill.

Implications. Difficulty with summatizing will obviously interfere with the students' progress in mathematics. Adding and subtracting will make little sense when students do not have a mental image of the numbers involved. Practicing summatizing is a quick process. Most often, choral drill is the easiest way to work with the entire class. Other times, students can work in smaller groups.

CORRECTION Modify these strategies according to students' learning needs.

1. *Choral Drill.* At the beginning of each class, get all the students' attention, and explain that you will show them a number of fingers for only 1 second. They are to call out the number of fingers you've shown. Show different numbers of fingers at random and move quickly. The entire drill ought to last no longer than 30–45 seconds. Be consistent in starting each class with this until most of the students can do it.

2. *Paired Drill.* After most of the students seem competent in summatizing, identify the few remaining who are still having difficulty. Pair each of those with a student who can run a private drill later during class or just before class begins. Remember, this drill should not last longer than 45 seconds.

3. *Expanded Choral.* The next step requires you to be a little nimble. Show the first set of fingers; students respond. Show the second set; again students respond. Then show the combined sum of the 2 sets and they should respond. At first, you may have to add a step between steps 2 and 3 where you say, "__ + __ = ?" and then show the total, which students call out.

4. *Extended Choral.* To transfer summatizing skills from fingers to other objects, use things that can be readily displayed (e.g., cards, straws, or pencils) or large pictures or transparencies of sets of objects. Follow the procedures of Activities 1, 2, or 3 to practice instant recognition of number value.

5. *Extended Practice.* Repeat the above activities many times. Use spiraling to go through each activity and then go back through all the activities; begin the process, move up, and then move back to the beginning. Many young learners need time to incorporate these notions. Bringing the notion up again and again after time for thinking is very helpful. This review of readiness skills will help make this basic learning permanent and central to the students' future progress.

17. ATTENDING TO SIGN

DETECTION This may be a special skill need of students who:

- Provide answers that make no sense using simple computation
- Read without stopping until the end of the line, regardless of punctuation
- Do not attend to details in problems
- Forget homework after they did it
- Seem to jump from one activity to another without closure

Description. Paying attention to the sign is placed here in the book because it is actually not a mathematics problem but a problem *for* mathematics. It affects every operation students attempt, as well as students' comprehension of problems in problem solving. Attending to the sign is simply looking at the symbol that tells you what operation is to be used to solve a problem. If inattention to sign is a behavioral problem, it is possible to treat it behaviorally.

Causation. There are three major causes of inattention to sign. First and most common, many boys and some girls enter school developmentally unready to attack the tasks set out for them. Their bodies are not ready to sit down and concentrate for long periods of time; they are still in the heavy-gross-motor movement stage. Second, some students have learned (often during the period just discussed) to attend incorrectly or to the inappropriate details; for example, they have learned to go directly to the ones column and therefore disregard the sign. Lastly, a small number of children are physiologically hyperactive. The cause of this is still not clearly known.

Implications. The implications for this difficulty are clear: If students do not attend to the sign and other details, they will fail in mathematics and many other subjects as well. They will not build the necessary foundation to move on in math and will miss new details as they are added to the process. For the first two groups, a strong behavioral program that emphasizes attention to detail is helpful. For the third group, a physician's help is needed, along with corrective strategies to overcome interfering habits.

CORRECTION Modify these strategies according to students' learning needs.

1. *Trace the Sign.* Have the student trace and read the sign 3 times before computing. This forces the student's attention directly to sign and away from other distracting factors.
2. *Match the Sign.* Put all 4 signs next to a series of problems and have the student draw lines between the corresponding problem and sign.
3. *Multiple Choice.* Take the sign out of each problem and put the answer in. Have the students determine which operation took place and fill in the appropriate sign.
4. *Color-Code.* Code each type of sign with a different color. Have students say the name of the color and the sign before computing. Have students who continue to experience difficulty color-code their own signs, again naming the color and sign before computation.
5. *Extended Practice.* Repeat the above activities many times. Use the spiraling process to review; rather than wait for mastery at one level, begin the process, move up, and then move back to the beginning. Many young learners need this extra review to incorporate these readiness skills.

18. USING DIRECTIONALITY CORRECTLY

DETECTION This may be a special skill need of students who:

- Compute in the incorrect direction
- Write numbers backward or reversed
- Move incorrectly while reading
- Add or subtract in the tens column first

Description.   Directionality is the structural way students approach computation. Although it is not the most critical area, in any way, it can cause students to fail in basic mathematics, which is cause for concern. Most school-oriented behavior is from left to right, and so it is not surprising that many students—even very bright students—simply apply this rule of movement to mathematics computation. In fact, when students are just beginning math and regrouping is not yet a factor, movement from left to right is not only normal but in fact works. The problem surfaces when regrouping is introduced about halfway through learning addition and subtraction. By this time, students have practiced this directional movement for such a long period of time that it may have become a genuine habit, one that could be hard to break. This is not to say that computation from left to right is mathematically incorrect but that the long-term effect of using that approach may be negative for many students with special needs.

Causation. Limited or inappropriate mathematics experience is the most likely cause of directionality problems. Some students may have ample experience with books and writing, both of which involve left-to-right movement. Thus, a general direction of interaction with the written symbol may have been established even before school entry. Some learning-disabled students have particular difficulties with directionality.

Implications. Correcting a directionality problem involves breaking a habit, which takes effort and concentration. Unfortunately, doing so means that the student cannot put all of his or her energy into new learning. Instead, both the teacher and the student must work to correct a problem that could have been dealt with much more easily early on in the process.

CORRECTION Modify these strategies according to students' learning needs.

1. *Color Columns.* Color-code the columns of the first sample problem on a given page. Use green and red for 2-column addition or subtraction. Use green, yellow, and red for 3-column addition or subtraction. Have the students color-code at least the first problem in each row so that they are an active force in this corrective measure.

2. *Dots.* Use dots at the top of each column to remind students that directionality must be addressed. Again the colors green, yellow, and red are familiar to students and convey a certain "go-stop" message. Have the students put the dots over problems other than the sample problem. Their interaction with the corrective process will help incorporate it into their learning system. Later just use a green dot over the ones place to show where to start.

3. *Arrows.* Draw an arrow across the top of the problem from right to left. (It is not critical that color be used; that is optional.) Have the students replicate the arrow above each problem. Later mark only the first problem in each row.

4. *Extended Practice.* Repeat the above activities many times using spiraling, where you go through each activity and then go back through all the activities. Many young learners need time to incorporate these notions; bringing them up again and again after time for thinking is very helpful.

REFLECTIONS

1. The organization of Part III suggests that the problems in mathematics readiness are different than those in higher mathematics (Part II). Compare and contrast a problem in each area. Are there distinct differences in the DETECTION behaviors and the CORRECTION strategies? Why or why not?

2. Much of the difficulty youngsters have with readiness is rooted in their limited development. Discuss this idea with others. See if you can identify students in kindergarten or first grade to whom this applies. Also discuss an alternative to current practices of student placement.

3. Many of the CORRECTION strategies apply to all students. Justify the selection of the ones presented, adding, deleting, or modifying strategies where you deem necessary.

4. Consider what might be done in a school if one-third to one-half of the students entering kindergarten were not ready to learn certain mathematics concepts. Select two skills from this chapter and explain.

5. Both teaching and learning mathematics are complicated processes. Based on the discussion of special mathematics needs in the previous chapters and your experience, for which readiness problems do you think regular classroom mathematics instruction is the most difficult to provide? Why? For which is it easiest? Why?

6. The CORRECTIVE PRINCIPLES in Part I are suggested as guides for selecting and modifying mathematics strategies. Select a hypothetical special learner; using the appropriate CORRECTIVE PRINCIPLES as guidelines, plan for that learner a modified basal mathematics lesson. Repeat the process for a second special learner. Compare and contrast the two lessons. For the same content, review the lesson script in the teacher's edition of a basal. How do your lessons differ from the ones suggested for most students?

7. Teachers often know how but do not have time to plan special mathematics lessons for special learners. Volunteer your services to plan one or more special mathematics readiness lessons for a special learner in a nearby classroom. Take the mathematics content of your lesson from the basal or other materials currently in use in that school. Use the diagnostic information available from the school and the CORRECTIVE PRINCIPLES to guide the design of your lesson.

8. Teachers also have trouble finding enough time to teach special mathematics lessons. Volunteer to actually teach the mathematics lessons you designed.

9. A number of mathematics and special education textbooks address the unique mathematics needs of special categories of students. Compare and contrast discussions in these sources with the information in Chapter 8:

Ashlock, R. B. (1986). *Error patterns in computation* (4th ed). Columbus, OH: Charles E. Merrill.

Choate, J. S., Bennett, T. Z., Enright, B. E., Miller, L. J., Poteet, J. A., & Rakes, T. A. (1987). *Assessing and programming basic curriculum skills.* Boston: Allyn and Bacon.

Heddens, J. W. (1984). *Today's mathematics.* Chicago: SRA.

Howell, K. W., & Morehead, M. K. (1987). *Curriculum based evaluation for special and remedial education.* Columbus, OH: Charles E. Merrill.

Johnson, S. W. (1979). *Arithmetic and learning disabilities.* Boston: Allyn and Bacon.

Reisman, F. K. (1982). *A guide to the diagnostic teaching of arithmetic.* Columbus, OH: Charles E. Merrill.

Schloss, P. S., & Sedlak, R. A. (1986). *Instructional methods for students with learning and behavior problems.* Boston: Allyn and Bacon.

PART IV

WHOLE-NUMBER COMPUTATION NEEDS

Whole-number computation deals with the various manipulations of whole numbers. Although there is no perfect definition of computation, it will be used here to include addition, subtraction, multiplication, and division and the basic number facts for those operations.

Computation will be described here as a three-part process. First, number facts are seen as the basic building blocks or foundations of computation. Without number-fact knowledge, there can be no computation; the learner will be bound to finger counting which, in terms of one-to-one correspondence, is really a readiness function. So number facts will be viewed as tools that must be memorized. Second, the computation skills for the four operations are also seen as tools to be used most effectively later in problem solving. However, the third function addressed has to do with the system of understanding that can be developed as computation skills are learned. This system of simple decision making and primary understanding of our number system serves the learner in higher mathematics processes such as problem solving.

Students develop concepts in many ways. As was pointed out earlier in this book, students must learn the concepts of computation within the context of problem solving. There has always been conflict between the academic purist and the practitioner. The purist believes that above all else the concept must be pure, or surely the student will be limited in his or her mathematics future. The practitioner believes that most special students need to concentrate on survival skills rather than reach for higher mathematics. Actually, both are right, and both are wrong. Some special students can achieve in higher mathematics, even though many cannot.

What should dictate the teaching approach is the manner in which a given student learns best. Not all learning has to take place from concept to application. In fact, emerging research indicates that some students learn more effectively by developing concepts through working with the structure of the algorithm. When possible, it is likely that we should start with the concept; however, if this is unsuccessful, we should not avoid other courses of action simply to maintain the purity of the skill.

So on the one hand, number facts and computation skills are viewed as tools, but on the other hand, they are seen as the means to begin the problem-solving (decision-making) process itself. They are critical tools for all learners but especially for learners with special needs. As these special-needs learners move through years of education, they will progress along the mathematics curriculum or fall further and further behind as a function of their fluency with these computation skills.

Part IV contains four chapters, each addressing one of the operational signs and related whole-number skills. Chapter 9 presents three special skill needs in the addition of whole numbers: producing 100 addition facts, regrouping, and applying an algorithm sequence. When working with special students in adding whole numbers, the teacher must make some difficult decisions regarding the cause of the inability. If a student shows no understanding of the function of addition, then the teacher is wise to begin with manipulatives and only much later move toward the actual mechanics of the skill. If, however, the student demonstrates adequate understanding and yet continues to answer addition questions incorrectly, then the teacher might want to try a step-by-step approach to addition. The implications of these problems—both to the instructional process and the students' later prospects—are also discussed in this section.

Chapter 10 presents three major skill areas in which special-needs students have difficulty when subtracting whole-number combinations. These skills parallel those examined in Chapter 9, but here they are explained as they relate to the subtraction of whole numbers. With regard to causation, there again seem to be two major approaches to consider. It appears that developmentally, students understand addition sooner than they understand subtraction. Therefore, it is possible that many more problems in subtraction are conceptual. Likewise, consider that even though subtraction is the inverse of addition, students may have problems in subtraction because of a weak understanding of addition. Thus, the reader might want to consider reviewing addition concepts before beginning any remedial program in subtraction. Again, implications for instruction and student progress are discussed, as well.

In Chapter 11, the three skill areas—producing facts, regrouping, and applying algorithms—are explained as they relate to the multiplication of whole numbers. Multiplication of whole numbers certainly separates successful students from unsuccessful students in computation. Since multiplication is built on the notion of repeated addition, students who had difficulty with addition will probably have similar problems with multiplication. It becomes even more complicated when students begin to multiply by numbers greater than 10. And students who previously experienced difficulty with basic set theory will have even more trouble when working with X number of sets of Y. The implications of these problems are also explained, both in terms of the instructional process and the students' later prospects in education if nothing is done about the problem.

Chapter 12 looks at the three major skill areas and learning difficulties in the division of whole numbers. Of all the areas of whole-number computation, division confuses the greatest number of learners. Controversy abounds about the problem of division and how to rectify it. Purists are opposed to using language like "goes into" because X does not really "go into" Y. Yet this may be the only way for certain students to develop any understanding of the process. Others may learn best by actually separating large groups of objects into equal-sized smaller groups. The point can only be repeated: What works best for one student may not work at all with another student. Each skill area is also discussed in terms of its implications for instruction and the students' educational prospects.

As mentioned earlier, there is a good deal of controversy over number facts, even how or if they should be taught. But the fact remains that there are clear differences on measures of achievement between students who know number facts and students who do not. Students who do not know the number facts operate more slowly and more often make careless or sloppy errors. And because they have to use some form of counting objects, time and attention are diverted away from higher-order skills such as decision making and problem solving and allocated to lower-level skills such as counting. Therefore, those who argue that we do not need math facts and that time spent there is wasted should realize that time is either invested in developing fundamental skills, or students must spend years compensating for the lack of those skills. This is even more the case for special-needs students, who typically need more time and practice than their peers.

Whole-number computation is the primary topic of the mathematics program for the first four years of formal mathematics training. Even so, many students fail to develop the skills and understanding necessary to progress to applying these skills to problem solving, fractions, and decimals. The activities suggested in Chapters 9 through 12 are proposed as a beginning to help students with special needs to catch up.

19. PRODUCING 100 NUMBER FACTS

DETECTION This may be a special skill need of students who:

- Do not seem to make predictable errors
- Count on their fingers
- Use a numberline for multidigit addition
- Frequently misstate simple addition answers

Description. There are 100 addition facts, which are those combinations of any two of all possible single-digit numbers that would result in a sum. Although these facts are frequently played down as being unimportant in this era of computers, they represent the foundation for all computation. Without knowing these facts, students are bound to a calculator or to counting on their fingers. To be sure students have committed the 100 addition facts to memory, check to see that all students can respond automatically when asked a number fact. If you use a written format, make sure the facts are out of order, and give students no more than three seconds per fact.

Causation. Limited auditory and associative memory can contribute to difficuties acquiring basic facts knowledge. Inadequate or insufficient practice of the basic facts often contributes to the problem, as does allowing students to continue to count on their fingers when they should have moved beyond that stage. Among the special categories of students who may have difficulties with this skill are those classified as learning disabled, mentally retarded, or language disabled.

Implications. The purpose for teaching and developing this skill is to lay the foundation for all future addition skills. Students who have problems learning the basic number facts will most likely have problems with all future mathematics operations. The relationship of the two single-digit numbers contained in the number fact is significant. Many beginning mathematics programs spend considerable time dealing with this conceptual relationship but expect students to translate this understanding into memorized facts without adequate repetition. The understanding of number relationships is not being questioned here but rather the transfer of that knowledge to facts.

CORRECTION Modify these strategies according to students' learning needs.

1. *Choral Drill.* Start each class with an oral practice session. This session should last no longer than 1-1/2 minutes. Select a group of about 30 facts each day. At first, you can use common facts (e.g., 1 + 1 = , 1 + 2 = , etc.); after a short time, you should mix your facts. State the fact in the form of a question (1 + 1 = ?) and have the class respond in unison. What will actually happen is that those who know the answer will respond instantly, while those who are learning will listen and repeat the answer almost instantly.

2. *Fact Test.* Check each student's basic fact acquisition using a written format. Mix all 100 addition facts in a horizontal format. This can be done on 2 sides of a single sheet. Have the students do side 1, giving them 150 seconds to do so. Tell the students not to use their fingers or any other device. Tell them to skip any fact that they do not automatically recall. Lastly, tell them it is important to complete this side of the paper even if they have to skip some of the problems. Check side 1. If there are 10 or more errors or omissions, do not give side 2.

3. *Fact Ring.* Using the results of Fact Test, select the first 10 facts each student either missed or skipped. Using small cards (3" x 2-1/2"), record the fact in question format on the front and the fact with the answer on the back. Punch a hole in the upper-left-hand corner of each card, and attach each student's set of cards on a ring that can be opened and closed. (A shower curtain ring works well for this.) Have the students practice using their rings when time permits during the day. They can check each other's progress in teams of 2. Once a fact has been mastered, take it off the ring and replace it with the next fact from the Fact Test.

4. *Speed Drill.* End each class with a 2-minute Speed Drill. This is much like Activity 2, except that students should continue to side 2 without stopping. Very few if any students will finish all the facts. However, students should be encouraged to try to do more facts each day. Students should correct their own papers, marking mistakes so that you can find any patterns to their errors.

5. *Fact Honor Roll.* Make a Fact Honor Roll that resembles a thermometer. Instead of degrees, put a blank line for each student's name. Laminate the poster so that you can reuse it each week. Use Friday's Speed Drill to determine which students have improved over the previous week. Place these students' names on the Fact Honor Roll.

20. REGROUPING

DETECTION This may be a special skill need of students who:

- Record the total answer for each column below the line
- Record the tens value below the line and carry the ones value
- Record the ones value and leave out the tens value
- Record the total sum below the line and also regroup
- Are unable to identify important place-value information

Description. Regrouping is a generic term that describes the exchange of value of a number across the appropriate place-value columns. For the purpose of addition, regrouping refers to breaking down a sum for a given column into its ones and tens values. The ones value must be placed in the ones column, and the tens value must be in the tens column. Regrouping is important for three reasons: 1) It demonstrates understanding of place value in basic numeration; 2) it is necessary to mechanically perform most addition skills; and 3) it is prerequisite to higher-level subtraction and multiplication.

Causation. There are two primary causes for difficulty with regrouping in the addition process. The first is conceptual weakness; that is, students often do not grasp the value of numbers and place value. This cause is serious because until remedied, students cannot be expected to build any further understanding into higher operations. The other cause of regrouping problems is more mechanical or structural. In this case, students simply learn the procedure incorrectly and use it without regard to the concept. That is, the procedure is formed outside of the area of understanding.

Implications. For students who exhibit special needs, problems with regrouping in addition of whole numbers will prevent them from learning all higher-level computational skills. Because there are two clearly different causes of this problem, no one solution will help all children. Students who have not developed the necessary numeration skills will need to review numeration, with heavy emphasis on the parts of numbers and their relative values. Students who show more structural deficits need not go so far back; their instruction should be based more on developing the correct procedures for regrouping. This means that the teacher will need to judge the cause of the difficulty as well as the difficulty itself. One way to do this is to have the students write numbers in their expanded forms as so many hundreds, so many tens, and so many ones. Inability to do this strongly suggests that they do not understand place value.

CORRECTION Modify these strategies according to students' learning needs. (Activities 3–5 are intended for students with more structural problems.)

1. *Manipulatives.* Use objects to help demonstrate the reality of numbers. If the number is 7, put 7 objects together. Students should be able to match these numbers to the corresponding sets of objects.

2. *Bundle Grouping.* Once students have successfully shown that they can match corresponding numbers and objects below 10, go beyond 10. Have the students bundle the objects into groups of 10 plus some number of objects left. Using boxes or cups labeled "ones" and "tens," have the students place the tens bundles into that box and the remaining loose objects into the ones box. On a piece of paper, label 2 columns "Ones" and "Tens," respectively. Have the students count the loose objects and record that value under the ones heading and count the bundles and record that under the tens heading. The number they have produced should equal the total number of objects.

3. *Red Box/Green Box.* Next to a sample problem, draw a green box and a red box to be used for the ones and tens values of column sums, respectively. When students add the ones column, they should record the sum for that column in the boxes; for example, if 8 + 9 are summed to produce 17, the 7 would go in the green box (ones) and the 1 would go in the red box (tens). Simultaneously, draw a green box under the ones column and a red box above the tens column; the 7 then moves from the green box next to the problem to the green box below the ones column (ones value to ones column), and the 1 in the red box next to the problem moves to the red box over the tens column (tens value to tens column). Structurally, this is an extension of the Bundle Grouping in Activity 2.

4. *Draw a Circle.* When Activity 3 seems to become almost automatic, eliminate the red and green boxes and draw 1 red circle above the tens column. This will serve the purpose of bridging between high structure and no structure. Far too often, corrective instruction moves drastically from high structure to no structure before real internal change has taken place in the students' thinking procedures.

5. *Extended Practice.* Provide enough structured and semistructured practice to replace the error with a correct procedure. How much practice is enough will vary from student to student, depending on how long the student has practiced regrouping in addition incorrectly and how well he or she understood the general functions of addition to begin with.

21. APPLYING ALGORITHM SEQUENCE

DETECTION This may be a special skill need of students who:

- Omit steps in a procedure
- Add using the multiplication process
- Place answers out of sequence
- Often get events out of sequence
- Utilize information incorrectly or repeatedly

Description. An algorithm is the sequence of events that should be followed in order to successfully complete a procedure. There is certainly more than one way to do addition skills. However, once a student has settled on a way, or algorithm, mixing up the sequence of that algorithm results in confusion and failure. Equally problematic is the development of faulty algorithms in addition. Algorithms build one upon another as addition skills become more complicated. Thus, when one defective algorithm develops, it naturally follows that later defects will be compounded.

Causation. Students who develop defective algorithms are often found to have had significant difficulties with sequencing events at an earlier age. These students have not genuinely developed an understanding of the cause/effect relationship. That is, they do not always see the relationship between the sequence of events and their effect on the outcome. These students are frequently disorganized in their approach to academics as well as their daily routine. Their notebooks often lack order; consequently, they frequently have trouble retrieving information needed to solve problems. Applying the multiplication process to addition demonstrates the student's urge to act quickly without regard to the effect of that behavior; often, the student has been practicing multiplication recently.

Implications. Students who develop defective algorithms in addition and also have difficulty with sequential functioning will have problems learning all higher-level math skills that require analytical thinking. This problem will also carry over into all other areas of the curriculum moreso than any other computation difficulty. Students with these types of difficulties should primarily work on developing sequential behaviors in arithmetic computation. This should not be confused with the school of thought that had students stringing beads 20 years ago in hopes that this would improve math and reading skills. Instead, this approach should emphasize sequence as a dynamic function of addition. All steps used to enhance sequential functioning should be presented in the context of addition of whole numbers. The most powerful means of assistance with sequence development in addition is the use of flowcharts as a sequential and self-monitoring guide.

CORRECTION Modify these strategies according to students' learning needs.

1. *Key Label.* To assist students in retrieving proper sequences for use within algorithms, have them make up titles or labels for the algorithm. As they are labeling the process, they are creating a means to call back the process through association.
2. *Problem Structure.* Diagram a sample problem for the students using lines, columns, arrows, or other tools that will clearly demonstrate a step-by-step procedure.
3. *Algorithm Flowchart:* **Regrouping in Addition**. Use the following flowchart to give students a visual picture of the total process and each step it contains. Have the students use a place holder (such as a coin) to help organize their movement through the algorithm.

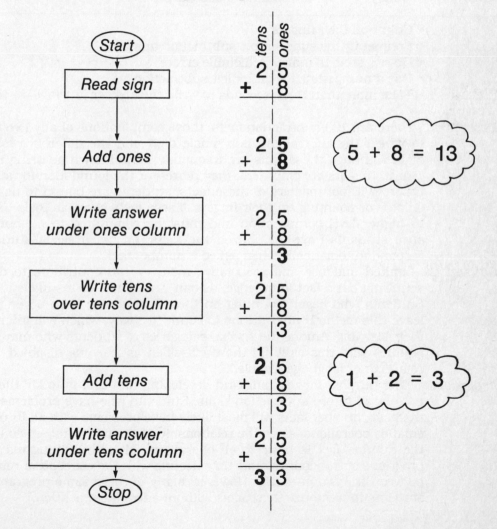

22. PRODUCING 100 NUMBER FACTS

DETECTION This may be a special skill need of students who:

- Count on their fingers
- Frequently misstate simple subtraction answers
- Do not seem to make predictable errors
- Use a numberline for multidigit subtraction
- Take more than three seconds to write the answer to a fact

Description. There are 100 subtraction facts, those combinations of any two numbers in which the subtrahend is a single digit and the difference is a single digit. Although these facts are frequently played down as being unimportant in this era of computers, they represent the foundation for all computation. Without mastery of these facts, students are bound to using a calculator or counting on their fingers. These facts serve as tools to be used in higher-level computation and must be committed to memory. To be sure students have memorized the facts, check to see if students can respond automatically when asked a number fact.

Causation. Limited auditory and associative memory can contribute to difficuties acquiring basic fact knowledge, as can inadequate or insufficient practice. Students who began counting on their fingers and were never taught to leave this method will continue to count this way, which will interfere with their learning. Among the special categories of students who may have difficulties with this skill are those classified as learning disabled, mentally retarded, or language disabled.

Implications. The purpose for teaching and developing this skill is to lay the foundation for all future subtraction skills. Students who have problems learning the basic number facts will most likely have problems with all future mathematics operations, since the relationship of the two numbers contained in the number fact is basic to all operations. Many beginning mathematics programs spend considerable time dealing with the conceptual relationship between the two numbers. However, many of these same programs expect students to memorize these facts without adequate repetition.

CORRECTION Modify these strategies according to students' learning needs.

1. *Choral Drill.* Start each class with an oral practice session. This session should last no longer than 1-1/2 minutes. Select a group of about 30 facts each day. At first, you can use common facts (e.g., 5 - 4 = , 5 - 3 = , etc.); after a short time, mix the facts. State the fact in the form of a question (5 - 4 = ?) and have the class respond in unison. What will actually happen is that those who know the answer will respond instantly, while those who are learning will listen and repeat the answer almost instantly.

2. *Fact Test.* Check each student's basic fact acquisition using a written format. Mix all 100 subtraction facts in a horizontal format. This can be done on 2 sides of a single sheet. Have the students do side 1, giving 150 seconds to complete this side. Tell the students not to use their fingers or any other device. Tell them to skip any fact that they do not automatically recall. Lastly, tell them it is important to complete this side of the paper, even if they have to skip some of the problems. Check side 1. If there are 10 or more errors or omissions, do not give side 2.

3. *Fact Ring.* Using the results of the Fact Test, select the first 10 facts each student either missed or skipped. Using small cards (3" x 2-1/2"), record the fact in question format on the front and the fact with the answer on the back. Punch a hole in the upper-left-hand corner of each card, and put each student's set of cards on a ring that can be opened and closed. (A shower curtain ring works well.) Have the students practice using their rings when time permits during the day. They can check each other's progress in teams of 2. Once a fact has been mastered, take it off the ring and replace it with the next fact from the Fact Test.

4. *Speed Drill.* End each class with a 2-minute speed drill. This is much like Activity 2, except that students should continue to side 2 without stopping. Very few if any students will finish all the facts. However, students should be encouraged to try to do more facts each day. Students should correct their own papers, marking mistakes so that you can find any patterns to their errors.

5. *Fact Honor Roll.* Make a Fact Honor Roll that resembles a thermometer. Instead of degrees, put a blank line for each student's name. Laminate the poster so that you can reuse it each week. Use Friday's Speed Drill to determine which students have improved over the previous week. Put these students' names on the Fact Honor Roll.

23. REGROUPING

DETECTION This may be a special skill need of students who:

- Take the smaller value from the greater, regardless of location
- Do not decrease the value of the number borrowed from
- Borrow when it is not necessary
- Regroup from the incorrect column
- Are unable to identify important place-value information

Description.   Regrouping is a generic term that describes the exchange of value of a number across the appropriate place-value columns. For the purpose of subtraction, regrouping refers to borrowing, or moving a set of tens into the ones column so that the value of the ones will equal or exceed the value of the subtrahend in that same column. Regrouping is important for three reasons: 1) It demonstrates whether the student understands place value in basic numeration; 2) it is necessary to operate mechanically with most subtraction skills; and 3) regrouping of subtraction is necessary to move on to division.

Causation.   There are two primary causes for difficulty with regrouping in subtraction. The first is conceptual weakness; that is, students often do not grasp the value of numbers and place value. This is a serious problem because until remedied, students cannot be expected to build any further understanding into higher operations. The other cause of regrouping problems is more mechanical or structural. In this case, students simply learn the procedure incorrectly and use it without regard to the concept. That is, the procedure is performed outside of the area of understanding.

Implications.   Special-needs students who have not mastered regrouping in subtraction of whole numbers will be unable to learn all higher-level computational skills. Because there are two clearly different causes of this problem, no one solution will help all children. Students who have not developed the necessary numeration skills will need to review numeration, with heavy emphasis on the parts of numbers and their relative values. Students who show more structural deficits need not go so far back; their instruction should be based more on developing the correct procedures for regrouping. This means that the teacher will need to judge the cause of the difficulty as well as the difficulty itself. One way to do this with regrouping is to have the students write numbers in their expanded forms as so many hundreds, so many tens, and so many ones. Students who cannot do this are most likely operating without knowledge of place value.

CORRECTION   Modify these strategies according to students' learning needs. (Activities 5–7 are intended for students with more structural problems.)

1. *Manipulatives.* Use objects to help demonstrate the reality of numbers. If the number is 7, put 7 objects together. Students should be able to match these numbers to the corresponding sets of objects.
2. *Greater or Less Than.* Using the typical format of subtraction problems, have the students compare the ones place of the subtrahend to the ones place of the minuend. The students should mark each problem so that they show which number is greater. This will lead to a more automatic behavior of checking the value of the numbers, a behavior that comes before regrouping.
3. *Star the Problem.* Create a poster that says, "If the bottom number is bigger, then you borrow," and put it up as a teaching poster. Have the students go back over all the problems they practiced in Activity 2 and star any problem in which the bottom number was bigger. These are the problems that require regrouping. Go on to have students star the numbers before attempting the problems on future practice pages.
4. *Bundle Breaking.* For students who seem to have difficulty understanding that they should add 10 and not 1 when borrowing, create tens bundles. Make a tens box and a ones box. Choose any number and put the correct number of tens and ones bundles in the appropriate boxes. Next take 1 bundle out of the tens box and add it to the ones box to show the regrouping. But first have the students break the band around the bundle to show there are 10 objects.
5. *Red Box/Green Box.* Above a sample problem, draw a green box and a red box to be used for regrouping a set of tens into the ones column (from green to red). For any problem where the bottom number is not greater than the top number, the students should draw a line through the boxes. For the other problems where regrouping is needed, have the students reduce the tens column by 1 and add that set of 10 to the value already in the ones column. This sum should be recorded in the red and green boxes. Subtraction should then continue as before.
6. *Draw a Circle.* When Activity 5 seems to become almost automatic, then just draw a circle above the tens column to continue to remind the students to reduce the value after they have borrowed. Many successful students become overly comfortable with regrouping and start to forget to reduce the value of the number borrowed from. This circle can be used to help remind them, too.
7. *Extended Practice.* Provide enough structured and semistructured practice to replace the error with a correct procedure. Many factors will influence the amount of practice each student needs, including how long he or she has practiced regrouping in subtraction incorrectly and how well he or she understood the general functions of subtraction to begin with.

24. APPLYING ALGORITHM SEQUENCE

DETECTION This may be a special skill need of students who:

- Omit steps in a procedure
- Subtract using the multiplication process
- Place answers out of sequence
- Often get events out of sequence
- Utilize information incorrectly or repeatedly

Description. An algorithm is the sequence of events that should be followed in order to successfully complete a procedure. There is certainly more than one way to do subtraction skills. However, once a student has settled on a way, or algorithm, mixing up that sequence results in confusion and failure. Equally as problematic is the development of defective algorithms in subtraction. Algorithms build one upon another as subtraction skills become more complicated. Thus, when students develop one defective algorithm, it naturally follows that later defects will be compounded.

Causation. Students who develop defective algorithms are often found to have had significant difficulties with sequencing events at an earlier age. These students do not genuinely understand cause/effect relationships. That is, they do not always see the relationship between the sequence of events and their effects on the outcome. These students are frequently disorganized in their approach to all school work. They do not develop predictable daily routines. Their notebooks often lack order; consequently, they frequently experience difficulty in retrieving information needed to solve problems. The mistake of applying the multiplication process to subtraction is made to provide an immediate solution to the problem without considering the effect of that behavior; often, the student has been practicing multiplication recently.

Implications. Students who develop defective algorithms in subtraction and also have difficulty with sequential functioning will have difficulty with all higher-level math skills that require analytical thinking. This will carry over into all other areas of the curriculum moreso than any other computation difficulty. Students with these types of difficulties should primarily work on developing sequential behaviors in arithmetic computation. Do not confuse this with the school of thought that had students stringing beads 20 years ago in hopes that this would teach them math and reading. Instead, this approach should emphasize sequence as a dynamic function of subtraction. All steps used to enhance sequential functioning should be presented in the context of subtraction of whole numbers. The most powerful means of assisting students with sequence development in subtraction is the use of flowcharts as sequential and self-monitoring guides.

CORRECTION Modify these strategies according to students' learning needs.

1. *Key Label.* To assist students in retrieving proper sequences for use within algorithms, have them make up titles or labels for the algorithm. As they are labeling the process, they are creating a means to call back the process through association.
2. *Problem Structure.* Diagram a sample problem for the students using lines, columns, arrows, or some other tools that will clearly demonstrate a step-by-step procedure.
3. *Algorithm Flowchart:* **Subtracting without Regrouping**. Use the following flowchart to provide students with a visual picture of the total process within which each step fits. Have the students use a place holder (such as a coin) to help organize their movement through the algorithm.

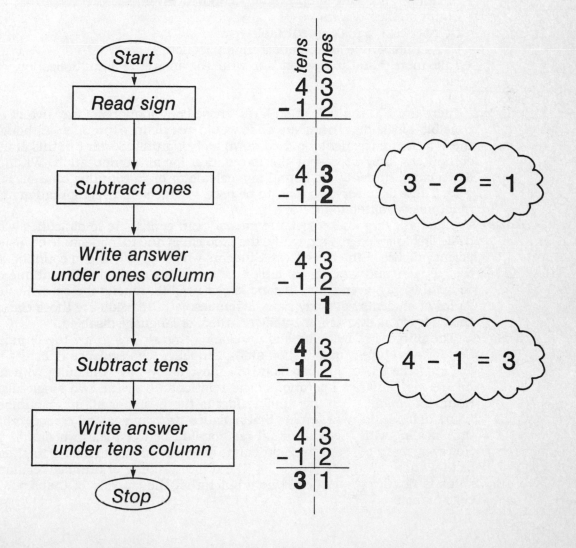

25. PRODUCING 100 NUMBER FACTS

DETECTION This may be a special skill need of students who:

- Frequently misstate simple multiplication answers
- Count on their fingers
- Do not seem to make predictable errors
- Use a numberline for multidigit multiplication
- Take more than three seconds to write the answer to a multiplication fact

Description. There are 100 multiplication facts, those combinations of any two of all possible single-digit numbers that would result in a product. Although these facts are frequently played down as being unimportant in this era of computers, they represent the foundation for all computation. Without these facts, students are bound to a calculator or to counting on their fingers. These facts serve as tools to be used in higher-level computation and must be committed to memory.

Causation. Limited auditory and associative memory can contribute to difficuties with basic fact knowledge. Frequently, the problem is due to inadequate or insufficient practice of the basic facts. Students who originally began counting on their fingers and were never taught to leave this method will continue to count this way, even when they no longer need to. Among the special categories of students who may have difficulties with this skill are those classified as learning disabled, mentally retarded, or language disabled.

Implications. The purpose of teaching and developing this skill is to lay the foundation for all future multiplication skills. Students who evidence difficulty in learning the basic number facts will most likely have difficulty with all future mathematics operations. The relationship of the two single-digit numbers contained in the number fact is the basis for all other number relationships. Many beginning mathematics programs spend considerable time dealing with the conceptual relationship between the two numbers. However, many of these same programs expect students to memorize these facts without adequate repetition. The understanding of number relationships is not being questioned here but rather the transfer of that knowledge to facts.

CORRECTION Modify these strategies according to students' learning needs.

1. *Choral Drill.* Start each class with an oral practice session. This session should last no longer than 1-1/2 minutes. Select a group of about 30 facts each day. At first, you can use common facts (e.g., 1 x 1 = , 1 x 2 = , etc.); after a short time, you should mix your facts. State the fact in the form of a question (2 x 2 = ?) and have the class respond in unison. What will actually happen is that those who know the answer will respond instantly, while those who are learning will listen and repeat the answer almost instantly.

2. *Fact Test.* Check each student's basic fact acquisition using a written format. Mix all 100 multiplication facts in a horizontal format. This can be done on 2 sides of a single sheet. Have the students do side 1, giving 150 seconds to complete this side. Tell the students not to use their fingers or any other device. Tell them to skip any fact that they do not automatically recall. Lastly, tell them it is important to complete this side of the paper, even if they have to skip some of the problems. Check side 1. If there are 10 or more errors or omissions, do not give side 2.

3. *Fact Ring.* Using the results of the Fact Test, select the first 10 facts each student either missed or skipped. Using small cards (3" x 2-1/2"), record the fact in question format on the front and the fact with the answer on the back. Punch a hole in the upper-left-hand corner of each card and attach each student's set of cards to a ring that can be opened and closed. (A shower curtain ring works well.) Have the students practice using their rings when time permits during the day. They can check each other's progress in teams of 2. Once a fact has been mastered, take it off the ring and replace it with the next fact from the Fact Test.

4. *Speed Drill.* End each class with a 2-minute Speed Drill. This is much like Activity 2 above, except that students should continue to side 2 without stopping. Very few if any students will finish all the facts. However, students should be encouraged to try to do more facts each day. Students should correct their own papers, marking mistakes so that you can find any patterns to their errors.

5. *Fact Honor Roll.* Make a Fact Honor Roll that resembles a thermometer. Instead of degrees, put a blank line for each student's name. Laminate the poster so that you can reuse it each week. Use Friday's Speed Drill to determine all students who have improved over the previous week. Place these students' names on the Fact Honor Roll.

26. REGROUPING

DETECTION This may be a special skill need of students who:

- Record the total product for each column separately below the line
- Record the tens value below the line and carry the ones value
- Record the ones value and leave out the tens value
- Are unable to identify important place-value information

Description. Regrouping is a generic term describing the exchange of value of a number across the appropriate place-value columns. For the purpose of multiplication, regrouping refers to breaking down a product for a given column into its ones and tens values. The ones value must be placed under the ones column, and the tens value should be incorporated into the next column. Regrouping is important for two reasons: 1) It demonstrates understanding of place value in basic numeration, and 2) it is necessary to mechanically operate with most multiplication skills.

Causation. There are two primary causes for difficulty with regrouping in the multiplication process. The first is conceptual weakness. That is, students often do not grasp the value of numbers and place value. This is a serious problem because until remedied, students cannot be expected to build any further understanding into higher operations. The other cause is more mechanical or structural. In this case, students simply learn the procedure incorrectly and use that procedure without regard to the concept; the procedure is performed outside of the area of understanding.

Implications. Students with special needs in the regrouping of whole numbers in multiplication will be effectively blocked from learning all higher-level computational skills. Because there are two clearly different causes of this problem, no one solution will help all children. Students who have not developed the necessary numeration skills will need to go back to numeration, with heavy review on the parts of numbers and their relative values. Students who show more structural deficits need not go so far back; their instruction should be based more on developing the correct procedures for regrouping. This means that the teacher will need to judge the cause of the difficulty as well as the difficulty itself. One way to do this with regrouping is to have the students write numbers in their expanded forms as so many hundreds, so many tens, and so many ones. Inability to do this strongly suggests that they are operating without knowledge of place value.

CORRECTION Modify these strategies according to students' learning needs. (Activities 3–5 are for students with more structural problems.)

1. *Manipulatives.* Use objects to help demonstrate the reality of numbers. If the number is 4, put 4 objects together. Students should be able to match these numbers to the corresponding sets of objects.

2. *Bundle Grouping.* Once students have successfully shown that they can match corresponding numbers and objects for numbers below 10, go beyond 10. Have the students bundle the objects into groups of 10 plus some number of objects left. Using boxes or cups labeled "ones" and "tens," have the students place the tens bundles into the tens box and the remaining loose objects into the ones box. On a piece of paper, label 2 columns "Ones" and "Tens," respectively. Have the students count the loose objects and record that value under the ones heading and count the bundles and record that under the tens heading. The number they have created should equal the total number of objects.

3. *Red Box/Green Box.* Next to a sample problem, draw a green box and a red box to be used for the ones and tens values of column sums, respectively. When students multiply the ones column, they should record the product for that column in the boxes (e.g., if the product of a column is 24, the 4 ones would go in the green box and the 2 tens would go in the red box). Simultaneously, draw a green box under the ones column and a red box above the tens column. The ones value (4) would then be moved from the green box next to the problem to the green box below the ones column (ones value to ones column). The tens value (2) in the red box next to the problem moves to the red box over the tens column (tens value into tens column). Structurally, this is an extension of the Bundle Grouping in Activity 2.

4. *Draw a Circle.* When Activity 3 seems to become almost automatic and the error of misplacing the parts of numbers is replaced with the correct procedure, eliminate the red and green boxes and replace them with 1 red circle above the tens column. This will bridge the gap between high structure and no structure. Far too often, corrective instruction moves drastically from high structure to no structure before real change has taken place in the students' thinking procedures.

5. *Extended Practice.* Provide enough structured and semistructured practice to correct the error. How much practice is enough will vary from student to student. The amount of practice needed will depend on how long the student has practiced regrouping in multiplication incorrectly and how well the student understood the general functions of multiplication to begin with.

27. APPLYING ALGORITHM SEQUENCE

DETECTION This may be a special skill need of students who:

- Multiply the ones column but add other columns
- Place answers out of sequence
- Often get events out of sequence
- Omit steps in a procedure
- Utilize information incorrectly or repeatedly
- Add using the multiplication process

Description. An algorithm is the sequence of events that should be followed in order to successfully complete a procedure. There is certainly more than one way to do multiplication skills. However, once a student has settled on a way, or algorithm, mixing up the sequence results in confusion and failure. The development of faulty algorithms in multiplication is equally as troublesome. Algorithms build one upon another as multiplication skills become more complicated. Thus, when one algorithm is defective, it naturally follows that later defects will be compounded.

Causation. Students who develop defective algorithms often are found to have had significant difficulty with sequencing events at an earlier age. These students do not understand cause/effect relationships. That is, they do not always see the relationship between the sequence of events and their effects on the outcome. These students are frequently disorganized in their approach to academics in general and also their daily routines. Their school work often lacks order; consequently, they often have difficulty retrieving information needed to solve problems. These students have difficulty correctly matching events with their appropriate counterparts. This problem occurs throughout the curriculum and thus is really a difficulty in the way these students approach learning in general.

Implications. Students who develop defective algorithms in multiplication and also have difficulty developing some sense of sequential functioning will have difficulty with all higher-level math skills that require analytical thinking. This will in fact carry over into all other areas of the curriculum, moreso than any other difficulty in computation. Students with these types of difficulties should primarily work on developing sequential behaviors in arithmetic computation. This should not be confused with the school of thought that once had students doing perfunctory skills unrelated to the academics they needed to develop. Instead, this approach emphasizes sequence as a dynamic function of multiplication. All steps used to enhance sequential functioning should be presented in the context of multiplication of whole numbers. The most powerful means of assisting students with sequence development in multiplication is the use of flowcharts as sequential and self-monitoring guides.

CORRECTION Modify these strategies according to students' learning needs.

1. *Key Label.* To assist students in retrieving proper sequences for use within algorithms, have them make up titles or labels for the algorithm. As they are labeling the process, they are creating a means to call back the process through association.
2. *Problem Structure.* Diagram a sample problem for the students using lines, columns, arrows, or some other tools that will clearly demonstrate a step-by-step means to proceed.
3. *Algorithm Flowchart:* **Regrouping in Multiplication**. Use the following flowchart to provide students with a visual picture of the total process within which each step fits. Have the students use a place holder (such as a coin) to help organize their movement through the algorithm.

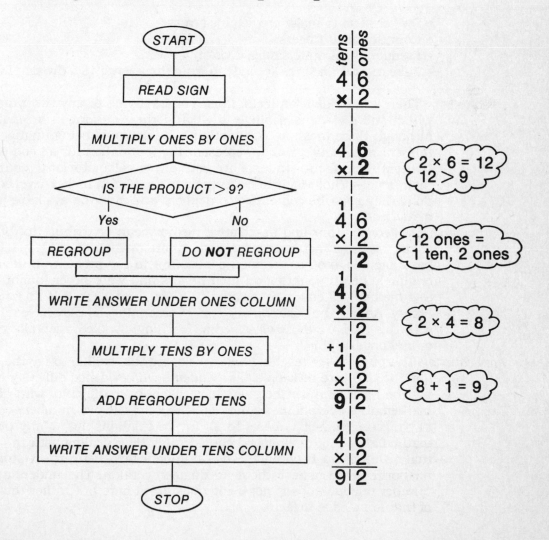

28. PRODUCING 90 DIVISION FACTS

DETECTION This may be a special skill need of students who:

- Do not seem to make predictable errors
- Count on their fingers
- Frequently misstate simple division answers
- Take more than three seconds to write the answer to a division fact

Description. There are 90 division facts, those combinations of any two numbers in which the divisor is a single digit and the quotient is a single digit. Although these facts are frequently played down as being unimportant in this era of computers, they represent the foundation for all computation. Without these facts, students are bound to a calculator or to counting on their fingers. These facts serve as tools to be used in higher-level computation. They must be committed to memory and must be available for automatic recall.

Causation. Limited auditory and associative memory can contribute to difficulties acquiring basic fact knowledge. Inadequate or insufficient practice of the basic facts is another common contributor to the problem. Students who originally began counting on their fingers and were never taught to leave this method will continue to count this way even when it is no longer necessary. Among the special categories of students who may have difficulties with this skill are those classified as learning disabled, mentally retarded, or language disabled.

Implications. The purpose for teaching and developing this skill is to lay the foundation for all future division skills. Students who evidence difficulty in learning the basic number facts will most likely have difficulty with all future mathematics operations, as the relationship of the two numbers contained in the number fact is basic to all future relationships. Many beginning mathematics programs spend considerable time dealing with the conceptual relationship between the two numbers but then expect students to memorize these facts without adequate repetition. The understanding of number relationships is not being questioned here but rather the transfer of that knowledge to facts.

CORRECTION Modify these strategies according to students' learning needs.

1. *Choral Drill.* Start each class with an oral practice session. This session should last no longer than 1-1/2 minutes. Select a group of about 30 facts each day. At first, you can use common facts (e.g., 6 + 2 =, 8 + 2 =, etc.); after a short time, mix the facts. State the fact in the form of a question (6 + 2 = ?) and have the class respond in unison. Actually, those students who know the answer will respond instantly, while those who are learning will listen and repeat the answer almost instantly.

2. *Fact Test.* Check each student's basic fact acquisition using a written format. Mix all 90 division facts in a horizontal format. This can be done on 2 sides of a single sheet. Have the students do side 1, giving 135 seconds to complete this side. Tell the students not to use their fingers or any other device. Tell them to skip any fact that they do not automatically recall. Lastly, tell them it is important to complete this side of the paper even if they have to skip some of the problems. Check side 1. If there are 10 or more errors or omissions, do not give side 2.

3. *Fact Ring.* Using the results of the Fact Test, select the first 10 facts each student either missed or skipped. Using small cards (3" x 2-1/2"), record the fact in question format on the front and the fact with the answer on the back. Punch a hole in the upper-left-hand corner of each card and attach each student's set of cards to a ring that can be opened and closed. (A shower curtain ring works well.) Have the students practice using their rings when time permits during the day. They can check each other's progress in teams of 2. Once a fact has been mastered, move it off the ring and replace it with the next fact from the Fact Test.

4. *Speed Drill.* End each class with a 2-minute Speed Drill. This is much like Activity 2 above, except that students should continue to side 2 without stopping. Very few if any students will finish all the facts. However, students should be encouraged to try to do more facts each day. Students should correct their own papers, marking mistakes so that you can find any patterns to their errors.

5. *Fact Honor Roll.* Make a Fact Honor Roll that resembles a thermometer. Instead of degrees, put a blank line for each student's name. Laminate the poster so that you can reuse it each week. Use Friday's Speed Drill to determine all students who have improved over the previous week. These students' names should be placed on the Fact Honor Roll.

29. REGROUPING

DETECTION This may be a special skill need of students who:

- Do not multiply or subtract for long division
- Are unable to identify important place-value information
- Just estimate how many times the divisor fits in each place of dividend and then round

Description. Regrouping is a generic term that deals with the exchange of value of a number across the appropriate place-value columns. For the purpose of division, regrouping refers to the steps in division after the student multiplies the partial quotient times the divisor to obtain a partial product. The regrouping then uses the difference generated added to the next place of the dividend, that step so often called "subtract and bring down." Students can miss this step by not subtracting or by not joining the difference to the next place of the dividend (the "bring down" part). This type of error can be easy to miss or misconstrue as a random fact error. Teachers will need to be alert to identify this error and realize that, unless found, the error will continue.

Causation. There are two primary causes of difficulty with regrouping in the division process. The first is conceptual weakness. That is, students often do not grasp the value of numbers and place value. This is a serious problem because until remedied, students cannot build any further understanding of higher operations. The other cause of difficulty is more mechanical or structural. In this case, students simply learn the procedure incorrectly and use it without regard to the concept. That is, the procedure is performed outside of the area of understanding.

Implications. Special needs with regrouping in the division of whole numbers will impede students' progression to higher-level computational skills. Because there are two clearly different causes of this problem, no one solution will help all children. Students who have not developed the necessary numeration skills will need to backtrack to numeration, with heavy emphasis on the parts of numbers and their relative values. Students who show more structural deficits need not go so far back; their instruction should center more on developing the correct procedures for regrouping. This means that the teacher will need to judge the cause of the difficulty as well as the difficulty itself. Emphasis should be placed on developing and then practicing regularly the skills of estimating in division. This is especially true for those with conceptual difficulty.

CORRECTION Modify these strategies according to students' learning needs. (Activities 3–5 address more structural problems.)

1. *Manipulatives.* Use objects to help demonstrate the nature of or readiness for estimating. Select any number (e.g., 25) as the dividend. Select a divisor that will divide equally into that other number (5). Then count out objects into groups equal to the divisor. Have students find how many groups it takes to equal the dividend. This is an ideal time to show students the relationship between division and multiplication.

2. *Bundle Grouping.* Build on Activity 1 by having the students bundle the divisor sticks. Again find how many bundles will be contained evenly in some larger number. Next, create a problem in the traditional format. Have the students try to divide the bundle into each place of the dividend, such as a bundle of 5 into a dividend of 65. The bundle goes into the 6 one time but not evenly; there is 1 left over. Ask the students, "Should we just throw away the extra 1?" Show how to subtract the 1 and add it to the next place of the dividend. This is useful in building a bridge between the use of manipulatives and the structure of the problem.

3. *Red Box/Green Box.* Under a sample problem, draw a green box and a red box to be used for the difference created when you subtract the partial product, which is generated by multiplying the partial quotient times the divisor and then subtracting it from the dividend. The red box should hold the difference and the green box should hold the next value of the dividend. Structurally, this is an extension of the Bundle Grouping in Activity 2.

4. *Draw a Circle.* When Activity 3 seems to become almost automatic and the error of misplacing the parts of numbers has been corrected, eliminate the red and green boxes and replace them with 1 red circle to hold the value brought down from each place of the dividend. This will bridge the gap between high structure and no structure. Far too often, corrective instruction moves drastically from high structure to no structure, before the students' understanding has really been established.

5. *Extended Practice.* Provide enough structured and semistructured practice to replace the error with a correct procedure. How much practice is enough practice will vary from student to student, depending on many factors. How long the student has practiced regrouping in division incorrectly and how well he or she understood the general functions of division to begin with are both important.

30. APPLYING ALGORITHM SEQUENCE

DETECTION This may be a special skill need of students who:

- Add the divisor to the dividend
- Subtract the divisor from the dividend
- Omit or do not label the remainder
- Place the answer out of sequence
- Often get events out of sequence
- Omit steps in a procedure
- Utilize information incorrectly or repeatedly

Description. An algorithm is the sequence of events that should be followed in order to successfully complete a procedure. There is certainly more than one way to do division skills. However, once a student has settled on a way, or algorithm, mixing up the sequence of that algorithm results in confusion and failure. The development of faulty algorithms in division is another serious problem. Algorithms build one upon another, as division skills become more complicated. Thus, when students develop one defective algorithm, it naturally follows that later defects will be compounded.

Causation. Students who develop defective algorithms are often found to have had significant difficulties with sequencing events at an earlier age. These students have not learned the value of having events occur in a fixed, sequential order; that is, they do not always see the relationship between the sequence of events and their effect on the outcome. These students are frequently disorganized in their approach to academics in general. They are also disorganized in their daily routine. Their homework often lacks order; consequently, they often have difficulty retrieving information needed to solve problems. Another error—applying the addition or subtraction process to division—is made because the student looks for an immediate solution to the problem without regard to the effect of that behavior.

Implications. Students who develop defective algorithms in division and also have sequential functioning problems will have difficulty with all higher-level math skills that require analytical thinking. This will in fact carry over into all other areas of the curriculum moreso than any other difficulty in computation. Students with these types of difficulties should primarily work on developing sequential behaviors in arithmetic computation. This does not relate to the school of thought that once had students stringing beads so they could learn math and reading. Instead, this approach emphasizes sequence as a dynamic function of division. All steps used to enhance sequential functioning should be presented in the context of division of whole numbers. The most powerful means to assist students with sequence development in division is the use of flowcharts as sequential and self-monitoring guides.

CORRECTION Modify these strategies according to students' learning needs.

1. *Key Label.* To assist students in retrieving proper sequences for use within algorithms, have them make up titles or labels for the algorithm. As they are labeling the process, they are creating a means to call it back through association.
2. *Problem Structure.* Diagram a sample problem for the students using lines, columns, arrows, or other tools that will clearly demonstrate a step-by-step procedure.
3. *Algorithm Flowchart:* **Division.** Use the following flowchart to provide students with a visual picture of the total process within which each step fits. Have the students use a place holder (such as a coin) to help organize their movement through the algorithm.

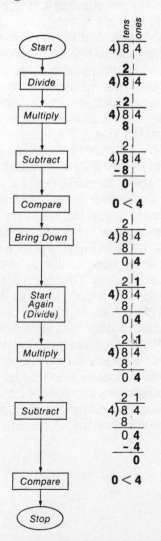

REFLECTIONS

1. The organization of Part IV suggests similiar types of difficulties in whole-number computation across the four basic operations. Compare and contrast a problem in each of the four operations. Are there distinct differences in the DETECTION behaviors and in the CORRECTION strategies? Why or why not?

2. Similar observable behaviors are cited for several skills; scan the DETECTION behaviors to locate commonalities across skills. Which particular skills cite the most similar behaviors? Why do you think this is so? Follow a similar procedure to compare CORRECTION strategies across skills.

3. Problems in mathematics tend to assume different proportions according to the student population and the perceptions of individual teachers. Interview a highly skilled regular education teacher to determine his or her perception of the important DETECTION behaviors and CORRECTION strategies for each categorized problem; then discuss detection and correction of any frequent problems that are not mentioned in Chapters 9–12. Follow a similar procedure to interview a veteran special education teacher.

4. Both teaching and learning mathematics are complicated processes. Based on the discussions of special mathematics needs in the previous chapters and your experience, for which problems in computation of whole numbers do you think regular classroom mathematics instruction is the most difficult to provide? Why? For which is it easiest? Why?

5. The CORRECTIVE PRINCIPLES in Part I are suggested as guides for selecting and modifying the CORRECTION strategies in Part IV. Select a hypothetical special learner; using as guidelines the appropriate CORRECTIVE PRINCIPLES, plan for that learner a modified basal mathematics lesson. Repeat the process for a second special learner. Compare and contrast the two lessons. For the same content, review the lesson script in the teacher's edition of a basal. How do your lessons differ from the ones suggested for most students?

6. Teachers often know how but don't have time to plan special mathematics lessons for special learners. Volunteer your services to plan one or more special mathematics lessons for a special learner in a nearby classroom. Take the mathematics content of your lesson from the basal or other materials currently in use in that school. Use the diagnostic information available from the school and the CORRECTIVE PRINCIPLES to guide the design of your lesson.

7. Teachers also have trouble finding enough time to teach special mathematics lessons. Volunteer to actually teach the mathematics lessons you designed.

8. A number of mathematics and special education textbooks address the mathematics needs of special categories of students. Compare and contrast discussions in these sources with the information in Chapters 9–12:

Ashlock, R. B. (1986). *Error patterns in computation* (4th ed.). Columbus, OH: Charles E. Merrill.

Choate, J. S., Bennett, T. Z., Enright, B. E., Miller, L. J., Poteet, J. A., & Rakes, T. A. (1987). *Assessing and programming basic curriculum skills.* Boston: Allyn and Bacon.

Enright, B. E. (1985). *ENRIGHT computation series* (Books A–D). N. Billerica, MA: Curriculum Associates.

Enright, B. E. (1983). *ENRIGHT diagnostic inventory of basic arithmetic skills.* N. Billerica, MA: Curriculum Associates.

Heddens, J. W. (1984). *Today's mathematics.* Chicago: SRA.

Howell, K. W., & Morehead, M. K. (1987). *Curriculum based evaluation for special and remedial education.* Columbus, OH: Charles E. Merrill.

Johnson, S. W. (1979). *Arithmetic and learning disabilities.* Boston: Allyn and Bacon.

Reisman, F. K. (1982). *A guide to the diagnostic teaching of arithmetic.* Columbus, OH: Charles E. Merrill.

PART V

FRACTION COMPUTATION NEEDS

Fractional numbers, as whole numbers, are a subset of the rational number system. They are most useful in describing the relationship between the number of parts an object is divided into and the number of those parts that are present or absent. This relationship is expressed in terms of X/Y, in which Y equals the number of parts a unit is divided into and X equals the number of equal parts that are being considered. The term *equal parts* is important because it underlies the very nature of fractional numbers.

Fractional numbers represent a precise equalness between each part. In the real world, we cannot divide objects perfectly into equal parts. However, we can approximate this operation, and for the purpose of several lessons in Part V that develop concept, we will do so. We will also restrict our discussion to common fractions, since operations of other types of fractional numbers are too advanced for our purposes.

Before students attempt to manipulate fractions, they must understand the basic nature of these numbers. For some reason, fractions are frequently viewed with great fear. This is strange because the students have been working with another subset of rational numbers for years (whole numbers). In order to understand the basic relationship of the parts of a fraction, students should be able to perform the four basic operations using whole numbers. They must at least understand the concepts behind those operations. For example, they must understand that division means breaking up a group into equal subgroups.

Next, the students can study the component parts of the fractional number itself. Young children understand the fraction one-half before they ever enter school, as they have been required to share objects by breaking them in half. But even though they know this concept, they frequently break the object (usually food) into two very unequal parts and then pick which "half" they want. In reality, there will never be perfect halves. However, this teaching process certainly helps develop the concept of equivalent parts.

The component parts of fractional numbers, as used in this book, are worth defining. The number of equivalent parts a group is divided into is determined by the *denominator* of the fraction. The denominator actually names the fraction and sets the rules for comparing fractions. The denominator is the part of the fraction that lies below the bar.

The *numerator* is that part of a fractional number that tells how many of the parts are being considered. That is, if the fraction is three-fourths, then there are four equal parts in the group and three of them are being considered. The numerator is the part of the fraction above the bar.

The *bar* or *fraction bar* is the line used to separate the numerator and the denominator. Later, in decimals and the conversion of fractions to decimals, the bar plays an even more significant role as the symbol for division.

Fractions can be demonstrated in a concrete, manipulative way. By dividing tangible objects into equivalent parts, students can see and feel the relationship between the parts and the whole. Fruit can be halved or quartered to show this relationship. Egg cartons are useful, especially since the parts are nearly perfectly equal in size; many activities can be planned, filling different numbers of parts to demonstrate different fractional numbers.

This type of simple activity can later be replaced by a more visually oriented model using a circle or a square. Choose a shape, draw in the number of parts (denominator), and shade in the number of parts being used (numerator).

Lastly, develop the writing of fractional numbers. Have the students practice writing fractions. Always start by having them write the denominator first, then the fraction bar, and lastly the numerator.

In Chapter 13, the addition of fractions is examined. In order for students to successfully master the skills in this chapter, as well as the next, stress must be placed on finding common denominators. Activities leading from manipulative orientation to process orientation are presented here.

Chapter 14 focuses on the skills needed in the subtraction of fractions. Students experience difficulties with two major concepts in this skill area: 1) the relationship of a whole object to its component parts and 2) the process of subtracting unlike fractions.

In Chapter 15, the multiplication of fractions is discussed. Although this is clearly the easiest mechanical set of skills in fractions, it does present some serious conceptual problems for special-needs students. Students who try to relate the multiplication of whole numbers with the multiplication of fractions often become confused, as the answers seem incongruent. Strategies are presented to assist with these special needs.

Chapter 16 looks at the special needs of students in the division of fractions. These skills present both conceptual and procedural difficulty for students. Moreover, students must have a clear understanding of the general division process before attacking fractions. The use of manipulatives is highly recommended for the development of these skills.

Fractional number computation is most strongly emphasized in grades four through six of the elementary curriculum. However, the basis for truly understanding fractions is developed in grades one through three. Students who try to learn these computation skills without the prerequisite understandings will most surely fail. And if nothing is done to provide the basic skills, these students will continue to have problems in mathematics and possibly other subjects, too.

31. DEMONSTRATING FRACTIONAL CONCEPTS

DETECTION This may be a special skill need of students who:

- Do not see a relationship between the parts and the whole
- Combine unlike fractions in addition
- Add the whole-number value of a mixed fraction to either the numerator or the denominator
- Cannot break a set into fractional equivalents
- Experienced difficulty understanding whole-number sets in addition or multiplication

Description. Fractional concepts are the fundamental facts of fractions. They can be as basic as the concept that there are fractions, that objects can be broken up into parts, and that at first those parts are equivalent. A child who has two pieces of candy can see how to share them between two people. However, how would that same child share them among four people? What is central to addition of fractions is that the fractional parts must be common in order to be joined easily. You can add one-half and one-fourth and get three-fourths, but that is only because you understand the relationship between fourths and halves. Thus, students must know that to add fractions, the denominator represents how many parts the whole is broken into, and the numerator represents how many of those parts are present. When you add two fractions, the number of parts the whole is broken into will not change, but the number of parts present will increase.

Causation. Students frequently have difficulties with fractional concepts in addition because they do not have a good understanding of whole-number sets. Many students are shown how to cut an apple in half, eat one-half, and find what is left; however, this simple example will not help students understand the relationship between denominators and numerators unless it is expanded greatly. Students who had significant difficulty with the concept of multiple equivalent sets for multiplication will certainly have problems with multiple equivalent partial sets.

Implications. Students who have difficulty with adding fractions will clearly have difficulty with higher-level fractional computation. They will also have difficulty with algebra and geometry. Before they spend significant time on the mechanics of adding fractions, students must develop a firm grasp of the fundamental relationship of the numerator to the denominator. This can be done in many ways and should be practiced over a long enough period of time so that it can become a basic part of students' mathematical understanding.

CORRECTION Modify these strategies according to students' learning needs.

1. *Split the Whole.* Put students into groups of 4. Take a selection of nutritional snacks, such as apples or oranges. Cut 1 piece of fruit in half and then in half again so that you have 4 quarters. Put the 4 equal pieces back together and wrap them with a rubber band. Hold up the rejoined piece of fruit and ask the group to tell you what you have (1 piece of fruit). Then take the rubber band off and ask if there is more fruit now. Show how the 4 equal pieces go together to make up the 1 whole piece of fruit. Let the students eat the fruit.

2. *Build the Denominator.* Building on Activity 1, explain to the group that the number of equal pieces something is cut into is called the *denominator.* Write that word on the board and have each student write it on a flashcard. Cut a piece of fruit in half. Ask, "How many pieces?" (2). Say, "We broke this fruit into 2 equal pieces; so if we wrote this as a fraction, the denominator would be 2." Cut the fruit into fourths and go through the process again. Take a banana and cut it into thirds and then sixths using the same procedure.

3. *Build the Numerator.* Building on Activities 1 and 2, explain to the group that "We know that the number of equal parts something is cut into is called the *denominator.* Now we need to find out how many of those parts we have; that is called the *numerator.*" Write that word on the board and have students write it on a flashcard. Cut an apple into fourths. Ask, "How many equal parts have we divided this apple into?" (4). Remind students that that is called the denominator. Give 1 part of fruit to someone to eat. Now ask, "How many of those 4 parts are left?" (3). Continue on, saying, "The number of parts that we have of something is the numerator, so the numerator here would be 3." Write the fraction 3/4 and review the relationship between the denominator and the numerator. Repeat this activity several times with varying fruits.

4. *Add the Fractions.* Once you have established the relationship between numerators and denominators, you can expand it into addition. Collect a batch of egg cartons. Cut up the egg-holder part, forming units of 2 cups, 3 cups, 4 cups, and so on. Have enough of each kind for every student. Get a bag of dried beans. Give each student a 2-cup egg holder. Ask, "How many equal parts are there?" (2). Say, "If this is divided into 2 equal parts, then the denominator for this fraction would be 2." Write it as a denominator on the board. Have each student take 1 bean and put it into 1 part of the holder. Ask, "How many parts of the egg carton are filled?" (1). Say, "If there could be 2 equal parts and we have filled 1 of those parts, the numerator is 1." Show the fraction 1/2. Pair two students to put their cartons next to each other. Say, "Now you have 1/2 and another 1/2. Put the bean from 1 carton into the empty space of the other carton. Now you have a carton with 2 equal parts; both of the parts are full. How could we write this as a fraction?" (2/2). Say, "So when we add like fractions together, we get larger fractions." Repeat this process for several days and then move to algorithms.

32. CONVERTING FRACTIONAL FORMS

DETECTION This may be a special skill need of students who:

- Leave answers in improper forms
- Add unlike fractions without first finding the lowest common denominator
- Do not reduce the fractional answer to its simplest form

Description. Converting fractions into a common form is a critical step in both the addition and subtraction of fractions. As the old saying goes, you can't add apples and oranges. In fractions, when you add 1/2 and 1/4 to get 3/4, you get that because you convert the 1/2 to 2/4 before you add. The same process holds exactly true for subtraction. If you do not convert before you add, you get 2/6 instead of 3/4. To convert to common form, the major step is to find the lowest common multiple into which all fractions being converted will divide evenly (lowest common denominator, or LCD). The other aspect of converting fractions is to present the answer in simplest or lowest form. The answer is still correct if it is not in simplest form; however, the simpler form is easier to work with and is more acceptable. An answer such as 6/12 may be correct but 1/2 is easier to see and work with.

Causation. Difficulty with converting fractions can be caused either by the students not knowing that all fractions can have various equivalent forms or by their not knowing the mechanics involved in determining the LCD. The first difficulty is a conceptual issue and the second is more of an operational defect. Both are common problems experienced by many types of special students.

Implications. Students who cannot convert fractions to common form simply cannot add or subtract fractions at all, which will bar them from most higher-level fraction problems. Students who cannot or do not reduce their answers can participate in most fractional problems but might produce unacceptable answers. If the cause of the difficulty is conceptual (i.e., students do not understand the relationships between the denominators of different fractions), experience with physically changing fractional forms will be necessary. Activity 1 and those listed for Special Need 31 are appropriate for correcting this problem. If the cause of the difficulty is operational, the use of a flowchart to find the LCD is recommended. Since this means of correction is also appropriate for subtraction, this will not be covered again under subtraction.

CORRECTION Modify these strategies according to students' learning needs.

1. *Convert the Form.* Take 2 pieces of construction paper, 1 red and 1 green. Draw a line across the red paper, creating 2 equal parts. Draw 3 lines across the green paper, creating 4 equal parts so that 2 of the equal parts of the green paper cover 1 part of the red paper. Cut the green paper into fourths, leaving the red paper intact. Label each part accordingly (1/2 or 1/4). Have the students cover a one-half portion of the red paper with 2 one-quarters. Stress the equalness of the 2 quarters and the half. This can be done with many different fractional forms.

2. *Algorithm Flowchart:* **Finding the LCD.** Use a flowchart that takes the students through each step in the procedure to find the LCD. One such flowchart is included below. You are encouraged to create an alternative flowchart for this process. Permit those who need it to continue to use the flowchart as a self-monitoring checklist.

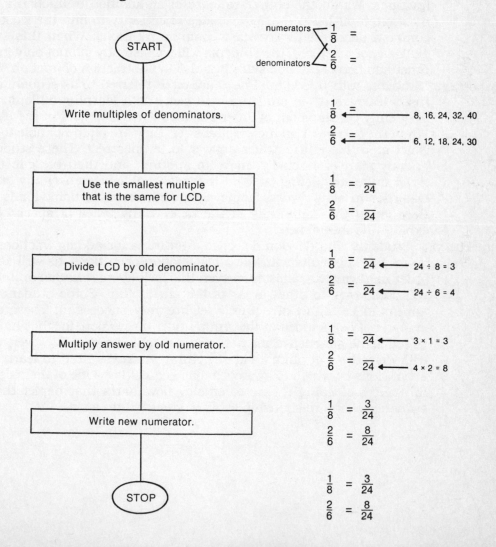

33. APPLYING ALGORITHM SEQUENCE

DETECTION This may be a special skill need of students who:

- Often get events out of order
- Omit steps in a procedure
- Utilize data incorrectly or repeatedly
- Multiply numerators
- Write the sum of the denominators
- Multiply denominators

Description. There is a sequence of at least 10 skills in the addition of fractions. To develop the ability to move from one skill to the next, the students will need to learn a correct algorithm. An algorithm is the sequence of events or steps that go together to successfully complete any one skill in adding fractions. Within the entire sequence of all addition of fraction skills, the most often mislearned are those steps necessary to find the least (lowest) common denominator (or least common multiple). When this subset of skills is not mastered, the students will consistently fail not only in higher-order addition of fraction skills but also in subtraction of fraction skills.

Causation. Students with this special need are often referred to as sequentially confused learners or as students with sequential learning difficulties. They frequently get steps out of order and also leave steps out of the algorithm. When they arrive at an incorrect answer, they are often not able to find the exact step that caused their answer to be incorrect. These students frequently start one activity, move to another, and then back to the first, never quite completing (at least in a reasonable sequence) any one event. Disorder in schoolwork, homework, and personal behavior is typical. Because of the complexity of the task, many types of special learners exhibit this special need.

Implications. Students who develop defective algorithms for adding fractions will no doubt have difficulty with the other fraction operations, as well as higher-order mathematics skills. Moreover, this lack of organization and structure will cross over to other areas of the curriculum. Some students do not appear to be highly structured yet are very successful. These students should not be included in the group under discussion, for they have established some alternative method of approaching skills. However, students with defective algorithms and structural weakness need to learn a useful set of steps by which to approach skill acquisition. One of the most powerful means of doing this is to employ flowcharts that depict the proper sequence of and relationship between steps within a skill.

CORRECTION Modify these strategies according to students' learning needs.

1. *Key Label.* To assist students in retrieving proper sequences for use within algorithms, have them make up titles or labels for the algorithm. As they are labeling the process, they are creating a means to call it back through association.
2. *Problem Structure.* Diagram a sample problem for the students using lines, columns, arrows, or some other tools that will clearly demonstrate a step-by-step procedure.
3. *Algorithm Flowchart:* **Adding Like Fractions.** Use the following flowchart to provide students with a visual picture of the total process within which each step fits. Have the students use some place holder (such as a coin) to help organize their movement through the algorithm. Permit students to use the chart as a self-monitoring checklist for as long as they need it.

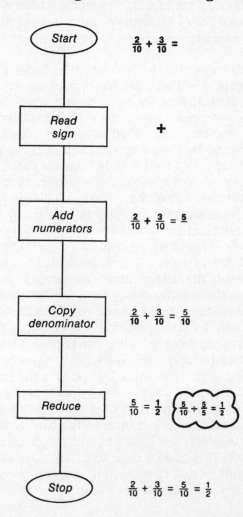

Start $\qquad \frac{2}{10} + \frac{3}{10} =$

Read sign $\qquad$ **+**

Add numerators $\qquad \frac{2}{10} + \frac{3}{10} = \frac{5}{}$

Copy denominator $\qquad \frac{2}{10} + \frac{3}{10} = \frac{5}{10}$

Reduce $\qquad \frac{5}{10} = \frac{1}{2} \quad \left\{ \frac{5}{10} \div \frac{5}{5} = \frac{1}{2} \right\}$

Stop $\qquad \frac{2}{10} + \frac{3}{10} = \frac{5}{10} = \frac{1}{2}$

34. DEMONSTRATING FRACTIONAL CONCEPTS

DETECTION This may be a special skill need of students who:

- Do not see a relationship between the parts and the whole
- Subtract unlike fractions
- Subtract the whole-number value of a mixed fraction from either the numerator or the denominator
- Cannot break a set into fractional equivalents
- Experienced difficulty understanding whole-number sets in subtraction or multiplication

Description. Fractional concepts, the fundamental facts of fractions, can be as basic as the concept that there are fractions, that objects can be broken up into parts, and that at first those parts are equivalent. The basic concepts in subtracting fractions have to do with separating parts of things to end up with some smaller value. What is central is that these parts must be common in order to be joined or separated easily. You can subtract 1/4 from 1/2 and get 1/4, but that is only because you understand the relationship between fourths and halves. Thus, students must know that to subtract fractions, the denominator represents how many parts the whole is broken into, and the numerator represents how many of those parts are present. When you subtract two fractions, the number of parts the whole is broken into will not change, but the number of parts present will decrease.

Causation. Students frequently have difficulties with fractional concepts in subtraction because they do not understand the concept of whole-number sets. Students must learn the relationship between denominators and numerators. Students who experienced significant difficulty with the concept of multiple equivalent sets for multiplication will certainly have greater problems with multiple equivalent partial sets. The concepts involved in the subtraction of fractions are especially difficult for many types of special learners.

Implications. Students who experience difficulty with the concepts in subtracting fractions will clearly have difficulty with higher-level fractional computation. They will also have difficulty with algebra and geometry. Before these students spend significant time on the mechanics of subtracting fractions, they need to grasp the fundamental relationship of the numerator to the denominator. This can be done in many ways and should be practiced over a long enough period of time so that it can become a basic part of the students' mathematical understanding.

CORRECTION Modify these strategies according to students' learning needs.

1. *Split the Whole.* Put students into groups of 4. Take a selection of nutritional snacks, such as apples or oranges. Cut 1 piece of fruit in half and then in half again so that you have 4 quarters. Put the 4 equal pieces back together and wrap them with a rubber band. Hold up the rejoined piece of fruit and ask the group to tell you what you have (1 piece of fruit). Now take the rubber band off. Ask the group if there is more fruit now. Show them how the 4 equal pieces go together to make up the 1 whole piece of fruit. Let the students eat the fruit.

2. *Build the Denominator.* Building on Activity 1, explain to the group how the number of equal pieces something is cut into is called the *denominator.* Write that word on the board and have each student write it on a flashcard. Cut a piece of fruit in half. Ask, "How many pieces are there?" (2). Say, "We broke this fruit into 2 equal pieces; so if we wrote this as a fraction, the denominator would be 2." Cut the fruit into fourths and go through the process again. Take a banana and cut it into thirds and then sixths using the same procedures.

3. *Build the Numerator.* Building on Activities 1 and 2, explain to the group that "We know that the number of equal parts something is cut into is called the *denominator.* Now we need to find out how many of those parts we have; that is called the *numerator.*" Write that word on the board and have students write it on a flashcard. Cut an apple into fourths. Ask, "How many equal parts have we divided this apple into?" (4). Remind students that that is called the denominator. Give 1 part to someone to eat. Now ask, "How many of those 4 parts are left?" (3). Say, "The number of parts that we have of something is the numerator, so the numerator here would be 3." Write the fraction 3/4 and review the relationship between the denominator and the numerator. Repeat this activity several times with varying fruits.

4. *Subtract the Fractions.* Once you have established the relationship between numerators and denominators, you can expand it into subtraction. Collect a batch of egg cartons. Cut up the egg-holder part, forming units of 2 cups, 3 cups, 4 cups, and so on. Have enough of each kind for every student. Get a bag of dried beans. Give each student a 2-cup egg holder. Ask, "How many equal parts are there?" (2). Say, "If this is divided into 2 equal parts, then the denominator for this fraction would be 2." Write it as a denominator on the board. Have each student take 2 beans and put 1 into each part of the holder. Ask, "How many parts of the egg carton are filled?" (2). Say, "If there could be 2 equal parts and we have filled 2 of those parts, the numerator is 2." Show the fraction 2/2. Have the students take 1 bean out and say, "You had 2/2 and you took away 1/2, or 1 bean. What is left? One bean is left. That means that when you have 2 equal parts and you remove 1 of those parts, you are left with 1/2. So when you have fractional parts and you take 1 or more of them away, you get a smaller fractional part." Repeat this process for several days and then move to the algorithm for subtraction.

35. APPLYING ALGORITHM SEQUENCE

DETECTION This may be a special skill need of students who:

- Often get events out of order
- Omit steps in a procedure
- Utilize data incorrectly or repeatedly
- Multiply numerators and denominators
- Write the sum of the denominators

Description. There is a sequence of at least nine skills in the subtraction of fractions. To be able to move from one skill to the next, students need to learn a correct algorithm form. An algorithm is the sequence of events or steps that go together to successfully complete any one skill in the subtraction of fractions. Within the entire sequence of all subtraction of fraction skills, the most often mislearned are those steps necessary to find the least (lowest) common denominator (or least common multiple). When this subset of skills is not mastered, the students will consistently fail not only in higher-order subtraction of fraction skills but also in other fraction skills.

Causation. Students with this special need are often referred to as sequentially confused learners or as students with sequential learning difficulties. These students frequently get steps out of order and also leave steps out of the algorithm. When they arrive at an incorrect answer, they are often not able to find the exact step that caused their answer to be incorrect. These students frequently start one activity, move to another, and then back to the first, never quite completing (at least in a reasonable sequence) any one event. Disorder in schoolwork, homework, and personal behavior is typical. Because of the complexity of the task, many special learners exhibit this special need.

Implications. Students who develop defective algorithms in subtracting fractions will no doubt have difficulty with the other operations in fractions and other higher-order mathematics skills. This lack of organization and structure will cross over to other areas of the curriculum as well. Some students do not appear to be highly structured yet are very successful. These students should not be included in the group under discussion, for they have established some alternative method of approaching skills. However, students with defective algorithms and structural weakness need to learn a useful set of steps by which to approach skill acquisition. The use of flowcharts that depict the proper sequence of steps within a skill, as well as the relationship between the steps, is a powerful means of correction.

CORRECTION Modify these strategies according to students' learning needs.

1. *Key Label.* To assist students in retrieving proper sequences for use within algorithms, have them make up titles or labels for the algorithm. As they are labeling the process, they are creating a means to call back the process through association.
2. *Problem Structure.* Diagram a sample problem for the students using lines, columns, arrows, or some other tools that will clearly demonstrate a step-by-step procedure.
3. *Algorithm Flowchart:* **Subtracting Like Fractions.** Use the following flowchart to give students a visual picture of the total process within which each step fits. Have the students use a place holder (such as a coin) to help organize their movement through the algorithm. Encourage students to use the chart for self-monitoring.

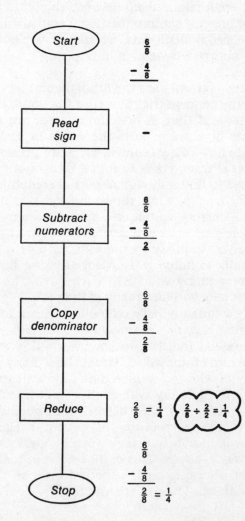

36. DEMONSTRATING FRACTIONAL CONCEPTS

DETECTION This may be a special skill need of students who:

- Do not see a relationship between the parts and the whole
- Cannot break a set into fractional equivalents
- Experienced difficulty understanding whole-number sets in multiplication
- Add or subtract instead of multiplying

Description. Fractional concepts are the fundamental facts of fractions. They can be as basic as the concept that there are fractions, that objects can be broken up into parts, and that at first those parts are equivalent. The denominator represents how many parts the whole is broken into, and the numerator represents how many of those parts are present. The basic concept in multiplication of fractions is to find a fractional part of something else. If students need to find a fractional part of something, their first inclination is to divide, for that is what they have learned in whole-number operations. Thus, the correct operation—multiplication—is confusing. The reality is that in fractions, first you divide to find what will be in each part or subset, and then you multiply by the number of those parts you wish to find. The denominator is actually a divisor showing how many parts you want and finding how many will be in each part. The numerator is the multiplier because it tells you how many of the partial sets you will end up with.

Causation. Students frequently have difficulties with fractional concepts in multiplication because they do not understand the concept of a fractional set. They do not understand that the denominator will give the number of equal groups, while the numerator will determine how many groups there are.

Implications. Students who experience difficulty with multiplication of fractions will clearly have difficulty with higher-level fractional computation. They will also have difficulty with algebra and geometry. Such students need to develop a firm grasp of the fundamental relationship of the numerator to the denominator before they spend significant time on the mechanics of multiplying fractions. This can be done in many ways and should be practiced over a long enough period of time so that it becomes basic to students' mathematical understanding.

CORRECTION Modify these strategies according to students' learning needs.

1. *Denominator as Divisor.* Give each student an egg carton and 24 counters, such as beans. Choose any number that is a multiple of the number of spaces in the given egg carton (if your carton has 6 spaces, pick 12, 18, or 24). Have the students divide the given number by the number of spaces in their carton. Have them put that number of beans into each space. Have them take out the beans from 1 space. These beans equal what $1/X$ would be. Repeat this several times each day for a couple of days.

2. *Numerator as Multiplier.* Building on Activity 1, once the denominator has been established as a divisor, have the students take more than 1 group of beans out. Write the fraction 3/6 on the board. Have students pick the egg carton with the appropriate number of spaces (6). Have them divide 12 beans equally among the 6 spaces. Ask the students to take out beans from 3 of the spaces, reminding them that the numerator tells how many sets to use. Tell them to keep the beans in groups of 2. They should have 3 groups of 2 on the desk. Ask the class to tell you 2 ways to find a total (add or multiply). Since the groups are equal, you can multiply. Have them add the beans and multiply to prove that both work. Point to the fraction 3/6 and say, "You divided 12 into 6 groups of 2; then you took 3 of those groups and you got 6. That is how 3/6 of 12 equals 6." Repeat this operation for several days. For the first few days, model the first example. Next pick a student to model an example. After about 5 days, quiz students on multiplying fractions. Let them use their egg cartons and beans. When you move to the algorithm for this skill, let them verify the algorithm using the manipulatives.

3. *Fractions of Fractions.* To show a fractional part of a fraction, follow the same general plan of Activities 1 and 2. However, use something that can be actually divided into parts. Cut an apple into fourths and separate them. The example will call for 1/2 times 1/4 or 1/2 of 1/4. Divide each of the fourths by the denominator 2. There are now 8 pieces separated into 4 groups; the fourths have become eighths. Take one of those groups (the 1/4). The numerator of the 1/2 tells you how many from each group you need (1). Take 1 out of the group and you end up with an eighth. Thus, 1/2 times 1/4 equals 1/8. Try it again with 2/3 times 3/4. Divide the apple into fourths and keep 3 of the fourths. Now divide each fourth by the denominator (3). You should now have 3 sets, each having three 1/12-size pieces of apples. The numerator of the first fraction tells us how many of those pieces we should take from each set (2). Take 2 of the pieces from each of the 3 sets and you have six 1/12-size pieces of apple. Thus, 2/3 times 3/4 equals 6/12. This can also be done with pieces of paper. Try this for several days; it will take a lot of practice to make this clear.

37. APPLYING ALGORITHM SEQUENCE

DETECTION This may be a special skill need of students who:

- Often get events out of order
- Omit steps in a procedure
- Utilize data incorrectly or repeatedly
- Write the sum of the denominators
- Use the common denominator

Description. There is a sequence of at least six skills in the multiplication of fractions. To be able to move from one skill to the next, the students need to learn a correct algorithm form, the sequence of events or steps that go together to successfully complete any one skill in the multiplication of fractions. Within the entire sequence of all multiplication of fraction skills, the two most often mislearned are 1) those steps necessary to put the answer into lowest or simplest form and 2) those steps used to convert mixed fractions into improper fractions so as to compute. When these subsets of skills are not mastered, the students will consistently fail not only in higher-order multiplication of fraction skills but also in division of fraction skills.

Causation. Students with these special needs are often referred to as sequentially confused learners or as students with sequential learning difficulties. They frequently get steps out of order and also leave steps out of the algorithm. When they arrive at an incorrect answer, they are often not able to find the step that caused their answer to be incorrect. These students frequently start one activity, move to another, and then back to the first, never quite completing (at least in a reasonable sequence) any one event. Disorder in schoolwork, homework, and personal behavior is typical.

Implications. Students who develop defective algorithms for multiplying fractions will have problems with the other operations in fractions and will likely develop serious problems with higher-order mathematics skills. Their inability to apply organization and structure will also affect other areas of the curriculum. Some students do not appear to be organized yet are very successful; they should not be included in this special group, for they have established some alternative method of approaching skills. However, students with defective algorithms and structural weakness need to learn a useful set of steps by which to approach skill acquisition. One of the most powerful means of doing this is the use of flowcharts that depict the proper sequence of steps within a skill, along with the relationship between the steps.

CORRECTION Modify these strategies according to students' learning needs.

1. *Key Label.* Have students make up titles or labels for each algorithm to assist them in retrieving proper sequences. As they are labeling the process, students are creating a means to call back the process through association.
2. *Problem Structure.* Diagram a sample problem for the students using lines, columns, arrows, or some other tools that will clearly demonstrate a step-by-step procedure.
3. *Algorithm Flowchart:* **Multiplying Like Fractions.** Use the following flowchart to provide students with a visual picture of the total process within which each step fits. Have students use a place holder (such as a coin) to help organize their movement through the algorithm. Continue the use of the chart as a self-monitoring aid as long as needed.

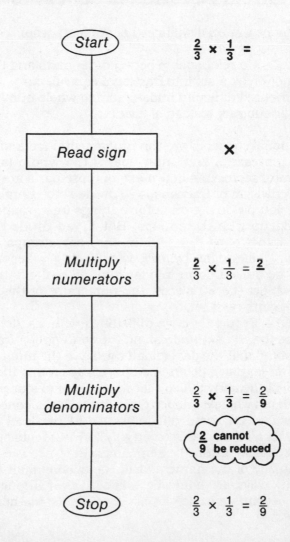

38. DEMONSTRATING FRACTIONAL CONCEPTS

DETECTION This may be a special skill need of students who:

- Do not see a relationship between the parts and the whole
- Cannot break a set into fractional equivalents
- Experienced difficulty understanding whole number sets in division
- Add or subtract instead of dividing

Description. Fractional concepts are the fundamental facts of fractions. The denominator represents how many parts the whole is broken into, and the numerator represents how many of those parts are present. The basic concept in division of fractions is to divide some number by a value less than one. If you divide a group of five things into groups of one (divide by one), you end up with five groups. But if you divide that same group of five things into groups of less than one—say, groups of 1/2—you need more groups. To divide five by, say, 1/4 (5 ÷ 1/4 =), each of the five objects first has to be broken down into fourths. All of the fourths are then added up. Once we get the 20 pieces, the numerator of the divisor will tell us how many go into each set.

Causation. Students frequently have difficulties with fractional concepts in division because they do not understand the concept of a fractional set. They do not understand that the denominator will give the number of equal groups while the numerator will determine how many groups there are. Division of fractions is particularly difficult for many types of students with special needs.

Implications. Students who experience difficulty with the concepts for division of fractions will clearly have difficulty with higher-level fractional computation, as well as algebra and geometry. Before students spend significant time on the mechanics of dividing fractions, they need to grasp the fundamental relationship of the numerator to the denominator. While this can be done in many ways, it should be practiced over a long enough period of time that it becomes a basic part of the students' mathematical understanding.

CORRECTION Modify these strategies according to students' learning needs.

1. *Denominator as Multiplier.* Example: 2 + 1/4 = __. Take 2 pieces of paper. Because the denominator (4) tells how many parts to break the whole into, cut each piece of the paper into 4 pieces. You now have 8 pieces. Practice this several times.

2. *Numerator as Divider.* After the students understand how the denominator determines how many pieces there will be, show them that you have kept the numerator at 1 and that this told how many pieces would be in each subgroup. That is, the 2 pieces were divided into groups of 1 one-fourths each. But what if the groups needed to have 3 one-fourths each (e.g., 9 + 3/4 =)? Each of the 9 pieces of paper would be cut into 4 equal parts, producing 36 parts. But the numerator (3) tells us how many pieces should be in each subgroup. Separating the 36 pieces of paper into groups of 3 gives us 12 groups. Thus, 9 pieces of paper divided into groups of three-fourths of a piece each will produce 12 groups of three-fourths piece each. Try this several times. Be sure that your example produces an even number of sets with no remainders. Guide students to develop similar examples.

3. *Divide Fraction by Fraction.* The same concept holds here: You take a fractional part and divide it into groups equal to the size of the divisor (e.g., 1/2 + 1/4 = ?). What you are essentially asking is, "How many one-fourths are there in one-half?" Take a whole apple and divide it into fourths. Since the example says that you do not start with a whole unit but only a half, get rid of a half. Now you have half an apple left; how many pieces do you actually have? (2). How many one-fourths are there in one-half? (2). Try another example: "How many sixths are there in two-thirds?" Cut the apple into sixths. Since you can only start with two-thirds, get rid of a third. Now count the number of sixths left (4). Try many of these examples. Use things that can be evenly cut up at first.

39. APPLYING ALGORITHM SEQUENCE

DETECTION This may be a special skill need of students who:

- Often get events out of order
- Omit steps in a procedure
- Utilize data incorrectly or repeatedly
- Write the sum of numerators and denominators
- Use the common denominator
- Do not convert mixed fractions to improper fractions

Description. There is a sequence of at least six skills in division of fractions. To be able to move from one skill to the next, the students need to learn a correct algorithm form. An algorithm is the sequence of events or steps that go together to successfully complete any one skill in the division of fractions. Within the entire sequence of division of fraction skills, the two most often mislearned sets of steps are those necessary 1) to put the answer into lowest or simplest form and 2) to convert mixed fractions into improper fractions so as to compute. When these subsets of skills are not mastered, students will consistently fail, both in higher-order division of fraction skills and other fraction skills.

Causation. Students with this special need are often referred to as sequentially confused learners or as students with sequential learning difficulties. They frequently get steps out of order and also leave steps out of the algorithm. When they arrive at an incorrect answer, they are often not able to find the step that caused the error. These students frequently start one activity, move to another, and then back to the first, never quite completing (at least in a reasonable sequence) any one event. Disorder in schoolwork, homework, and personal behavior is typical. Division of fractions is often the most difficult operation for most special learners.

Implications. Students who develop defective algorithms in the division of fractions will no doubt have difficulty with the other operations in fractions and will likely develop serious difficulties in higher-order mathematics skills. This lack of organization and structure will cross over to other areas of the curriculum as well. Some students do not appear to be highly structured yet are very successful. They should not be included in the group under discussion, for they have established some alternative method of approaching skills. However, students with defective algorithms and structural weakness need to learn a useful set of steps by which to approach skill acquisition. One of the most powerful means of doing this is the use of flowcharts that depict the proper sequence of and relationship between events within a skill.

CORRECTION Modify these strategies according to students' learning needs.

1. *Key Label.* To assist students in retrieving proper sequences for use within algorithms, have them make up titles or labels for the algorithm. Doing this will provide a means to call back the process through association.
2. *Problem Structure.* Diagram a sample problem for the students using lines, columns, arrows, or some other tools that will clearly demonstrate a step-by-step procedure.
3. *Algorithm Flowchart:* **Dividing Like Fractions.** Use the following flowchart to provide students with a visual picture of the total process within which each step fits. Have the students use a place holder (such as a coin) to help organize their movement through the algorithm. Encourage students to use the chart as a self-monitoring aid until the skill is learned.

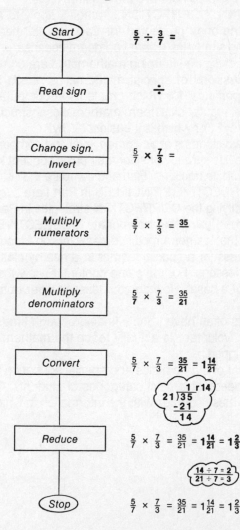

REFLECTIONS

1. The organization of Part V suggests similar types of problems in the computation of fractional numbers across different operations. Compare and contrast a problem in each area. Are there distinct differences in the DETECTION behaviors and the CORRECTION strategies? Why or why not?

2. Select one operation of fractions and design a lesson to introduce it. Be sure to include the use of manipulatives and the transfer to a visual model. Determine all prerequisite knowledge students will need, and create a quick pretest to assess these skills.

3. Find a class that would be appropriate to try your pretest with. Make sure to give the test enough in advance to either change your lesson or prepare those students who do not have the required skills.

4. Many of the CORRECTION strategies apply to all students. Justify the selection of the ones presented for each special needs area, adding, deleting, or modifying strategies where you deem necessary.

5. Both teaching and learning mathematics are complicated processes. Based on the discussions of special mathematics needs in the previous chapters and your experience, for which problems in computation of fractional numbers do you think regular classroom mathematics instruction is the most difficult to provide? Why? For which is it easiest? Why?

6. Select two students at any grade level and compare their knowledge of fractional numbers. Be sure to consider their basic concept knowledge as well as their ability to compute fractions. Find and compare their strengths and weaknesses.

7. The CORRECTIVE PRINCIPLES in Part I are suggested as guides for selecting and modifying the CORRECTION strategies in Part V. Select a hypothetical special learner; using the appropriate CORRECTIVE PRINCIPLES as guidelines, plan for that learner a modified basal lesson for computation of fractions. Repeat the process for a special learner in a nearby classroom. Compare and contrast the two lessons. For the same content, review the lesson script in the teacher's edition of a basal. How do your lessons differ from the ones suggested for most students?

8. Teachers often have trouble finding enough time to teach special mathematics lessons. Volunteer to actually teach the mathematics lessons on fractions that you designed.

9. A number of mathematics and special education textbooks address the mathematics needs of special categories of students. Compare and contrast discussions in these sources with the information in Chapters 13–16:

Ashlock, R. B. (1986). *Error patterns in computation* (4th ed.). Columbus, OH: Charles E. Merrill.

Choate, J. S., Bennett, T. Z., Enright, B. E., Miller, L. J., Poteet, J. A., & Rakes, T. A. (1987). *Assessing and programming basic curriculum skills.* Boston: Allyn and Bacon.

Enright, B. E. (1985). *ENRIGHT computation series* (Books E–I). N. Billerica, MA: Curriculum Associates.

Enright, B. E. (1983). *ENRIGHT diagnostic inventory of basic arithmetic skills.* N. Billerica, MA: Curriculum Associates.

Heddens, J. W. (1984). *Today's mathematics.* Chicago: SRA.

Howell, K. W., & Morehead, M. K. (1987). *Curriculum based evaluation for special and remedial education.* Columbus, OH: Charles E. Merrill.

Johnson, S. W. (1979). *Arithmetic and learning disabilities.* Boston: Allyn and Bacon.

Reisman, F. K. (1982). *A guide to the diagnostic teaching of arithmetic.* Columbus, OH: Charles E. Merrill.

PART VI

DECIMAL COMPUTATION NEEDS

Decimal computation deals with the various manipulations of decimal numbers, including addition, subtraction, multiplication, and division. Decimal computation will be addressed in this part as both a complete process and a tool to use later in higher math. As a process, decimals help students learn the relationship between parts and wholes.

The beauty of decimals in comparison to fractions is that decimals stay with base ten. In addition, decimals are used in all international measurement systems, namely, the metric system. If one develops the concepts of base ten when learning whole-number concepts, then learning decimals is the logical extension of that system when dealing with parts of wholes. Frequently, people speak of fractions and how it is easier to think in fractions than to think in decimals. This may be because those people are used to fractions.

Students must be exposed to decimals not just as a mechanical process but in terms of the relationship of decimal operations to operations already mastered. If students have had significant difficulty with the computation of whole numbers and that difficulty still exists, it is doubtful that they will benefit from instruction in decimals.

A major component of computation is the tying together of the uses of whole numbers, fractions, and decimals. This tying together brings closure to the computation process. It should expand the students' understanding of the number system and help clarify the important relationships that exist between whole numbers, fractions, and decimals.

For students to move on to calculating percent, interest, or discount, decimals and their operations must be both understood and mastered. And as students move up in mathematics to algebra and geometry, they will need to employ whole numbers, decimals, and fractions with flexibility. This can only occur after the student both understands and can manipulate these numbers.

Chapter 17 examines the special needs students experience when learning to add decimals. As with the subtraction of decimals, students may be confused about the relationship of tenths to hundredths. The confusion often results in the misalignment of numbers.

Chapter 18 investigates the skills associated with the subtraction of decimals. Here, as in Chapter 17, students must have a clear understanding of the relationship of the parts to the whole that was developed in fractions. Students again experience more difficulty in crossing over the decimal than they do in the mechanical part of subtraction.

Chapter 19 presents the special needs of students experiencing difficulty with multiplying decimals. Many special-needs students either leave out the decimal point or place it incorrectly. If the student is proficient in the mechanics of multiplying whole numbers, stress should be placed on the flowcharts presented to teach how to place the decimal point in multiplication.

Chapter 20 examines the difficulties special-needs students have when learning to divide decimals. Special care should be taken to first check the students' ability with the division of whole numbers. If this proves adequate, then use the activities provided to help students understand the division of decimals.

Decimal numbers becomes an important part of the basal curriculum between fourth and seventh grade, depending on which basal is used. The difficulties many students have directly result from their poor understanding of the relationship between the whole-number computation skills they already know and the decimal skills being introduced. The activities of Part VI stress the use of manipulatives to reintroduce students to this relationship.

40. DEMONSTRATING PLACE-VALUE UNDERSTANDING

DETECTION This may be a special skill need of students who:

- Leave out the decimal point
- Add numbers without regard to column values
- Record their answers without regard to decimal values
- Cannot show the relationship of the parts to the whole

Description. The concept of place value for the addition of decimals is a direct outgrowth of that for the addition of whole numbers. Students must extend their understanding of the value of each column to the right of the decimal point as well as to the left of the decimal point. Thus, students who had difficulty grasping the concept of place value when adding whole numbers will most likely have difficulty understanding place value when adding decimals. As with fractions, students must see the relationship of the parts and the whole. The advantage to working with decimals is that the entire system is set up in base ten. With fractions, the base is constantly changing as a function of the denominator, whereas with decimals, it remains constant and can be related to the whole-number value. One concept that is particularly difficult for some students is that just because a number has more digits, it may not be a larger number. With whole numbers, if a number has an extra digit (432 instead of 32), the number is automatically greater. With decimals, that is not necessarily the case; it is the value of the tenths place, not the length of the decimal, that most often determines relative value.

Causation. The most obvious cause for difficulty with place-value concepts in the addition of decimals is difficulty with place-value concepts in the addition of whole numbers. Another common problem is students' difficulty visualizing tenths or hundredths of objects. The concept of breaking whole objects into partial states will be difficult for students with special learning needs to master.

Implications. The previous statement should not be taken to mean that students with special learning needs cannot learn to add decimals. If that were the case, they would not be able to use decimals at all. The point is that there is no practical way to use numbers if you do not understand their function and basic concepts. Special-needs students will simply need more experience using decimal parts.

CORRECTION Modify these strategies according to students' learning needs.

1. *Fractional Review.* Review with students the relationship of the parts and the whole they learned in fractions. Ask them to name items that can be divided into equal parts. See how many the class can name.

2. *Paper Tenths.* Take a strip of paper 10 inches long. Have the students divide the paper into 10 1-inch sections. Have them turn the paper over and write the words "1 piece" on the opposite side. Ask the students how many pieces of paper they have (1). Then have them turn it over and ask how many parts the paper is divided into (10). Tell them in decimals, these are called tenths. Ask them to shade in or color 4 of the tenths. Write the number .4 on the board.

3. *Paper Strips.* Give each student 2 more paper strips already divided and labeled. Ask the students to show you 3 tenths by covering 3 tenths of 1 piece of the paper. Use a few more examples. Now have the students show you 1 whole and 3 tenths. They should hold up the 1 strip and cover 3 sections of the other. Explain that 10 tenths is the same as 1 whole strip or the number 1. Write the number 1.5 on the board and ask them to show you that value.

4. *Decimal Chart.* Put a decimal place-value chart on the board.

ones	.	tenths
1	.	4
3	.	5

    Read the following numbers to the class and, after modeling the first 2–3, have individual students fill in the place-value chart: 1 & 4 tenths; 3 & 5 tenths; 2 & 7 tenths; 5 & 2 tenths; 8 & 9 tenths; 4 & 1 tenth; etc.

5. *Counting Graph.* Using graph paper, give each student a square with 100 blocks in it, 10 x 10. On the reverse side, students should write "1 piece." Have them turn over the paper and then explain that something can be broken down into 100 pieces as easily as 10 pieces; these parts are called hundredths. Have the students cover 5 hundredths. Next have them cover 10 hundredths, then 15 hundredths, etc. Then go back and write those numbers on the board: .05 = 5 hundredths, etc. Write several additional hundredths values on the board and have students cover the appropriate amounts.

6. *Decimal Chart II.* Put a place-value chart on the board again for tens, ones, tenths, and hundredths. Read a series of numbers containing the above parts and have students fill in the parts.

41. APPLYING ALGORITHM SEQUENCE

DETECTION This may be a special skill need of students who:

- Often get events out of order
- Omit steps in a procedure
- Utilize data incorrectly or repeatedly
- Multiply instead of add
- Add using the multiplication process
- Omit or misplace the decimal point

Description. A sequence of at least 11 skills is involved in the addition of decimals. To be able to move from one skill to the next, students need to learn a correct algorithm form, that sequence of events or steps that go together to successfully complete any one skill in the addition of decimals. The most common errors made in applying algorithms for the addition of decimals are the same as those made in the addition of whole numbers. In fact, students who demonstrated errors such as those described in Special Needs 19–21 and who did not receive appropriate corrective instruction have a high probability of making the same errors in adding decimals. In addition to those errors, students frequently omit the decimal point from their answer. Although this may seem minor, teachers must stress the difference between .250 and 250 in a real-life situation, such as writing a check.

Causation. Students who exhibit this special need are often described as being sequentially confused or as having sequential learning difficulties. They frequently get steps out of order and also leave steps out of the algorithm. Thus, when they arrive at an incorrect answer, they are usually unable to find the problem. These students frequently start one activity and move to another, alternating and never quite completing (at least in a reasonable sequence) any one event. Disorder in schoolwork, homework, and personal behavior is typical.

Implications. Students who develop defective algorithms in the addition of decimals will no doubt have difficulty with other decimal operations and will likely develop difficulty in higher-order mathematics skills. This lack of organization and structure will affect other areas of the curriculum as well. Some students do not appear to be highly structured yet are very successful; they should not be included in the group under discussion, for they have established some alternative method of approaching skills. However, students with defective algorithms and structural weakness need to learn a useful set of steps by which to approach skill acquisition. One of the most powerful means of doing this is by employing flowcharts that depict the proper sequence of events within a skill along with the relationship between the steps.

CORRECTION Modify these strategies according to students' learning needs.

1. *Key Label.* Have students make up titles or labels for the algorithm to assist them in retrieving proper sequences. As they label the process, they are creating a means to call it back through association.
2. *Problem Structure.* Diagram a sample problem for the students using lines, columns, arrows, or some other tools that will clearly demonstrate a step-by-step means to proceed.
3. *Algorithm Flowchart:* **Adding Decimals.** Use the following flowchart to provide students with a visual picture of the total process within which each step fits. Have the students use a place holder (such as a coin) to help organize their movement through the algorithm. Encourage the continued use of the chart for self-monitoring until learned.

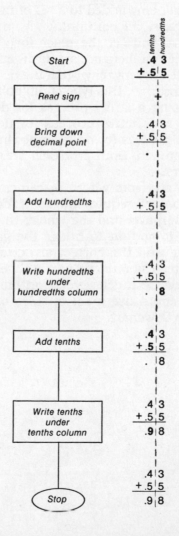

42. PLACING THE DECIMAL POINT

DETECTION This may be a special skill need of students who:

- Reverse the position of digits in the answer
- Omit or misplace the decimal point
- Place the answer too far to the left or right of the decimal
- Add but use the multiplication rule to place the decimal

Description. Placement of the decimal point in adding decimals is a structural process. Problems with this can also reflect the lack of place-value understanding. It is necessary to differentiate between these two types of problems. The first will be dealt with in this section; the second was dealt with in Special Need 40. The decimal point must be placed with regard to the largest decimal being added to a set of numbers. That is, if you are adding several numbers, all containing decimal parts, the number with the longest decimal will set the place for the decimal point in the answer. Normally, if the numbers are arranged in column form, the solution is easier because the student may bring the decimal down.

Causation. As mentioned earlier, this type of difficulty can be either conceptual or the result of learning a faulty process for decimal placement. Students who have learned a defective process for decimal placement have probably practiced that process for quite some time and will be consistent in their particular approach. Both problems are often exhibited by many types of special learners.

Implications. In general, students whose placement problems are a result of place-value confusion will require intense work with manipulatives to grasp the meaning of the parts and the whole; this was dealt with in Special Need 40. It will be important to bridge the gap between this new place-value understanding and the correct structural procedure for decimal placement. And after the student has reached that understanding, do not let him or her develop a defective algorithm. To develop a correct algorithm, the students could use a set of rules or that set of rules could be diagrammed by a flowchart.

CORRECTION Modify these strategies according to students' learning needs.

1. *Concept Review.* Review the CORRECTION strategies for Special Need 40 with the students. Make sure they have a firm grasp of the concepts of the whole and its parts.
2. *Addition of Decimals.* Use the following set of steps to help the students develop the sequence for correctly placing the decimal in the addition of decimals. Practice using these steps with the problems below.

   a. Make sure that all the columns are lined up in the proper order. All the decimal points should be in a straight vertical line.
   b. Add the numbers in each column, using regrouping when necessary.
   c. Bring the decimal point straight down into the answer.
   d. Count the places to the right of the decimal point for each addend.
   e. Take the largest single answer from step d and count from the decimal point in the answer to the right to the end of the answer. The number of spaces to the right of the decimal in the answer should equal the largest single number of places in step d.

1)	2)	3)	4)
21.5	32.45	25.745	13.7
+ 34.3	+ 11.23	+ 44.253	+ 22.456

   5) 22.4 + 16.55 =

   6) 57.62 + 21.259 =

3. *Extended Practice.* Once you have modeled the appropriate steps for students, have them work in groups of 2, creating problems and answering them. This can be expanded into a class game of Placing the Point, where teams challenge each other with their problems.

43. DEMONSTRATING PLACE-VALUE UNDERSTANDING

DETECTION This may be a special skill need of students who:

- Leave out the decimal point
- Subtract the numbers without regard to the column values
- Record their answers without regard to decimal values
- Cannot show the relationship of the parts to the whole

Description. The concept of place value for the subtraction of decimals is a direct outgrowth of that for the subtraction of whole numbers. Students must extend their understanding of the value of each column to the right as well as to the left of the decimal point. Students who had difficulty grasping the concept of place value in the subtraction of whole numbers will almost surely have difficulty with understanding place value in the subtraction of decimals. As with fractions, students must see the relationship of the parts and the whole. The advantage to working with decimals is that the entire system is set up in base ten. With fractions, the base is constantly changing as a function of the denominator. With decimals, it remains constant and can be related to the whole-number value. One concept that is particularly difficult for some students is that just because a number has more digits, it may not be a larger number. With whole numbers, if a number has an extra digit (432 instead of 32), it is automatically greater. With decimals, that is not necessarily the case; it is the value of the tenths place, not the length of the decimal, that most often determines relative value.

Causation. The most obvious cause of difficulty with place-value concepts in the subtraction of decimals is a poor understanding of place-value concepts in the subtraction of whole numbers. Students also often have difficulty visualizing tenths or hundredths of objects.

Implications. The previous statement should not be taken to mean that students with special learning needs cannot learn how to subtract decimals. If that were the case, they would not be able to use decimals at all. There is no practical way to use numbers if you do not understand their function and basic concepts. Instead, the implication is that special-needs students will need more experience using decimal parts.

CORRECTION Modify these strategies according to students' learning needs.

1. *Fraction Review.* Review with the students the relationship of the parts and whole they learned in fractions. Ask them to name items that can be divided into equal parts. See how many the class can name.
2. *Paper Tenths.* Take a strip of paper 10 inches long. Have the students divide the paper into 10 1-inch sections. Have them turn the paper over and write the words "1 piece" on the opposite side. Ask students how many pieces of paper they have (1). Then have them turn it over and ask how many parts the paper is divided into (10). Tell them in decimals, these are called tenths. Ask them to shade in or color 4 of the tenths. Write the number .4 on the board.
3. *Paper Strips.* Give each student 2 more strips already divided up and labeled. Ask the students to show you 3 tenths by covering 3 tenths of 1 piece of the paper. Use a few more examples. Now have the students show you 1 whole and 3 tenths. They should hold up the 1 strip and cover 3 sections of the other. Explain that 10 tenths is the same as 1 whole strip or the number 1. Write the number 1.5 on the board and ask them to show you that value.
4. *Decimal Chart.* Put a decimal place-value chart on the board.

ones	.	tenths
2	.	5
4	.	7

   Read the following numbers to the class and after modeling the first 2–3, have individual students fill in the place-value chart: 2 & 5 tenths; 4 & 7 tenths; 3 & 9 tenths; 6 & 4 tenths; 8 & 2 tenths; 5 & 4 tenths; etc.
5. *Counting Graph.* Using graph paper, give each student a square with 100 blocks in it, 10 x 10. On the reverse side, they should write "1 piece." Have them turn over the paper and explain that something can be broken down into 100 pieces as easily as 10 pieces and that these parts are called hundredths. Have the students cover 5 hundredths. Next have them cover 10 hundredths, then 15 hundredths, etc. Then go back and write those numbers on the board: .05 = 5 hundredths, etc. Write several additional hundredths values on the board and have students cover the appropriate amounts.
6. *Decimal Chart II.* Put a place-value chart on the board again for tens, ones, tenths, and hundredths. Read a series of numbers containing the above parts and have the students fill in the chart.

44. APPLYING ALGORITHM SEQUENCE

DETECTION This may be a special skill need of students who:

- Often get events out of order
- Omit steps in a procedure
- Utilize data incorrectly or repeatedly
- Multiply instead of subtract
- Subtract using the multiplication process
- Omit or misplace the decimal point

Description. There is a sequence of at least 11 skills in the subtraction of decimals. To be able to move from one skill to the next, the students will need to learn a correct algorithm form. An algorithm is the sequence of events or steps that go together to successfully complete any one skill in the subtraction of decimals. The most common errors made in subtraction of decimals algorithms are the same as those made in subtraction of whole numbers algorithms. In fact, students who demonstrated errors such as those described in Special Needs 22–24 and who did not receive appropriate corrective instruction will most likely make the same errors in subtracting decimals. In addition to those errors, students frequently omit the decimal point from their answers.

Causation. Students who exhibit this special need are often referred to as sequentially confused learners or as students with sequential learning difficulties. They frequently get steps out of order and also leave steps out of the algorithm. When they arrive at an incorrect answer, they usually cannot find the exact step that caused their answer to be incorrect. These students frequently start one activity, move to another, and then back to the first, never quite completing anything (at least in a reasonable sequence).

Implications. Students who develop defective algorithms in the subtraction of decimals will no doubt have difficulty with the other operations in decimals; they will likely develop serious difficulties in higher-order mathematics skills as well. Moreover, this lack of organization and structure will cross over to other areas of the curriculum. Some students do not appear to be highly structured yet are very successful. These students should not be included in the group under discussion, for they have established some alternative method of approaching skills. However, students with defective algorithms and structural weakness need to learn a useful set of steps by which to approach skill acquisition. One of the most powerful means of doing this is to use flowcharts that depict the proper sequence of events within a skill.

CORRECTION Modify these strategies according to students' learning needs.

1. *Key Label.* To assist students in retrieving proper sequences for use within algorithms, have them make up titles or labels for the algorithm. Labeling the process creates a means to call it back through association.
2. *Problem Structure.* Diagram a sample problem for the students using lines, columns, arrows, or some other tools that will clearly demonstrate a step-by-step procedure.
3. *Algorithm Flowchart:* **Subtracting Decimals.** Use the following flowchart to provide students with a visual picture of the total process within which each step fits. Have the students use a place holder (such as a coin) to help organize their movement through the algorithm. Until this skill is mastered, have students use the chart as a self-monitoring aid.

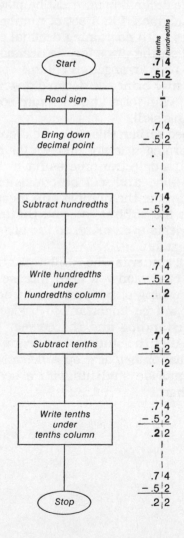

45. PLACING THE DECIMAL POINT

DETECTION This may be a special skill need of students who:

- Reverse the position of digits in their answer
- Omit or misplace the decimal point
- Place the answer too far to the left or right of the decimal
- Subtract but use the multiplication rule to place the decimal

Description. Placement of the decimal point in the subtraction of decimals is a structural process. Difficulty can reflect the lack of place-value understanding. It is necessary to differentiate between these two types of problems. The first will be dealt with in this section; the second was dealt with in Special Need 43. The decimal point must be placed with regard to the largest decimal being subtracted in a set of numbers. That is, if you are subtracting two numbers, both containing decimal parts, the number with the longest decimal will set the place for the decimal point in the answer. Normally, if the numbers are arranged in column form, the solution is easier because the student may bring the decimal down. However, math problems are not always in column form and therefore some rules must exist to assist students with this skill.

Causation. As mentioned earlier, this type of difficulty can be either conceptual or the result of learning a faulty process for decimal placement. Students who have learned a defective process have probably practiced that process for quite some time and will be consistent in their particular approach. Students may line the numbers up incorrectly and thus the decimal point will move in the subtrahend and the minuend. This type of mistake will confuse students even more, so the alignment of the problem is an important consideration.

Implications. In general, students who exhibit placement problems that are a result of place-value confusion will need intense experience with manipulatives to grasp the meaning of the parts and whole. This was dealt with in Special Need 43. It will be important to bridge the gap between the new place-value understanding and the correct structural procedure for decimal placement. After the student has acquired the place-value understanding, do not let him or her develop a defective algorithm. To develop a correct procedure, provide students with a set of rules or summarize the rules into a flowchart.

CORRECTION Modify these strategies according to students' learning needs.

1. *Concept Review.* Review the exercises in Special Need 43 with the students. Make sure they have a firm grasp of the concepts of the whole and its parts.
2. *Subtraction of Decimals.* Use the following set of steps to help the students develop the sequence for correctly placing the decimal in the subtraction of decimals. Practice using these steps with the problems below.

   a. Make sure that all the columns are lined up in the proper order. All the decimal points should be in a straight vertical line.
   b. Subtract the numbers in each column, using regrouping when necessary.
   c. Bring the decimal point straight down into the answer.
   d. Count the places to the right of the decimal point for both numbers.
   e. Take the largest single answer from step d and count from the decimal point in the answer right to the end of the answer. The number of spaces to the right of the decimal in the answer should equal the largest single number of places in step d.

1)	2)	3)	4)
21.5	32.45	25.745	13.7
— 10.4	— 21.23	— 3.62	— 9.45

5) 22.4 — 16.55 =

6) 57.62 — 21.259 =

3. *Extended Practice.* Once you have modeled the appropriate steps above for the students, have them work in groups of 2, creating problems and answering them. This can be expanded into a class game of Placing the Point, where teams challenge each other with their problems.

46. DEMONSTRATING PLACE-VALUE UNDERSTANDING

DETECTION This may be a special skill need of students who:

- Leave out the decimal point
- Multiply the numbers without regard to the column values
- Record their answers without regard to decimal values
- Cannot show the relationship of the parts to the whole

Description. The concept of place value for the multiplication of decimals is a direct outgrowth of that for the multiplication of whole numbers. Students must extend their understanding of the value of each column to the right of the decimal point as well as to the left of the decimal point. Students who had difficulty grasping the concept of place value for the multiplication of whole numbers will almost surely have difficulty understanding place value for the multiplication of decimals. As with fractions, students must see the relationship of the parts and the whole. The advantage to working with decimals is that the entire system is set up in base ten. With fractions, the base is constantly changing as a function of the denominator, whereas with decimals, it remains constant and can be related to the whole-number value. One concept that is particularly difficult for some students is that just because a number has more digits, it may not be a larger number. With whole numbers, if a number has an extra digit (432 instead of 32), the number is automatically greater. With decimals, that is not necessarily the case; it is the value of the tenths place, not the length of the decimal, that most often determines relative value.

Causation. The most obvious cause for difficulty with place-value concepts in the multiplication of decimals is difficulty with place-value concepts in the multiplication of whole numbers. Students also often have difficulty visualizing tenths or hundredths of objects. The concept of breaking down whole objects into partial states will be difficult for students with special learning needs.

Implications. The previous statement should not be taken to mean that students with special learning needs cannot learn to multiply decimals. If that were the case, they would not be able to use decimals at all. There is no practical way to use numbers if you do not understand their function and basic concepts. Instead, the implication is that the students will need more experience using decimal parts.

CORRECTION Modify these strategies according to students' learning needs.

1. *Fraction Review.* Review the relationship of the parts and the whole students learned in fractions. Ask students to name items that can be divided into equal parts. See how many the class can name.
2. *Decimal Chart.* Put a decimal place-value chart on the board.

ones	.	tenths
2	.	7
4	.	5

Read a list of numbers to the class and after modeling the first 2–3, have individual students fill in the place-value chart.
3. *Counting Graph.* Using graph paper, give each student a square with 100 blocks in it, 10 x 10. On the reverse side, students should write "1 piece." Have them turn over the paper and explain that something can be broken down into 100 pieces as easily as 10 pieces and that these parts are called hundredths. Have the students cover 5 hundredths. Next have them cover 10 hundredths, then 15 hundredths, etc. Then go back and write those numbers on the board: .05 = 5 hundredths, etc. Write several additional hundredths values on the board and have students cover the appropriate amounts.
4. *Decimal Chart II.* Put a place-value chart on the board again for tens, ones, tenths, and hundredths. Read a series of numbers containing the above parts and have the students fill in the parts.

Example: 25.78

tens	ones	.	tenths	hundreths
		.		

5. *Graph-Paper Multiplication.* Using the graph paper, have students cover 5 of the hundredths. Have them cover 2 more groups of 5 hundredths each. Have them count how many are covered. Have them write their answer in decimal form. Now go back and discuss what has just taken place as a function of multiplication. That is, students have taken 3 groups of 5 hundredths and produced 15 hundredths, or .15. Focus on the fact that the reason you could multiply is the same for decimals as it is for whole numbers: The groups were of equal size.

47. APPLYING ALGORITHM SEQUENCE

DETECTION This may be a special skill need of students who:

- Often get events out of order
- Omit steps in a procedure
- Utilize data incorrectly or repeatedly
- Add instead of multiply in different columns
- Omit or misplace the decimal point

Description. The multiplication of decimals involves a sequence of at least six skills. To develop the ability to move from one skill to the next, students need to learn a correct algorithm form, the sequence of events or steps that go together to successfully complete any one skill in the multiplication of decimals. The most common errors made in applying algorithms for the multiplication of decimals are the same as those made in the multiplication of whole numbers. In fact, students who demonstrated errors such as those described for Special Needs 25–27 and who did not receive appropriate corrective instruction have a high probability of making the same errors in multiplying decimals. In addition to those errors, students frequently omit the decimal point from their answers.

Causation. Students who exhibit this special need are often referred to as being sequentially confused or having sequential learning difficulties. They frequently get steps out of order, leave steps out of the algorithm, and when they arrive at an incorrect answer, are unable to find the problem. These students frequently start one activity, move to another, and then back to the first, never quite completing (at least in a reasonable sequence) any one event.

Implications. Students who develop defective algorithms in the multiplication of decimals will no doubt have difficulty with the other operations in decimals and will likely develop serious difficulties in higher-order mathematics skills. This lack of organization and structure will cross over to other areas of the curriculum as well. Some students do not appear to be highly structured yet are very successful; they should not be included in the group under discussion, for they have established some alternative method of approaching skills. However, students with defective algorithms and structural weakness need to learn a sequential approach to skill acquisition. One of the most powerful means of doing this is to use flowcharts that depict the proper sequence of events within a skill.

CORRECTION Modify these strategies according to students' learning needs.

1. *Key Label.* To assist students in retrieving proper sequences for use within algorithms, have them make up titles or labels for the algorithm. As they are labeling the process, they are creating a means to call back the process through association.
2. *Problem Structure.* Diagram a sample problem for the students using lines, columns, arrows, or some other tools that will clearly demonstrate a step-by-step procedure.
3. *Algorithm Flowchart:* **Multiplying Decimals.** Use the following flowchart to provide students with a visual picture of the total process within which each step fits. Have the students use a place holder (such as a coin) to help organize their movement through the algorithm. Encourage students to use the chart as a self-monitoring aid as needed.

$$\begin{array}{r} 4.79 \\ \times 2.34 \end{array}$$

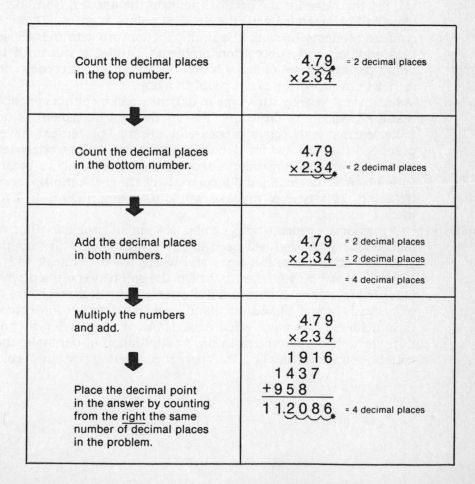

Count the decimal places in the top number.	4.79 = 2 decimal places ×2.34
Count the decimal places in the bottom number.	4.79 ×2.34 = 2 decimal places
Add the decimal places in both numbers.	4.79 = 2 decimal places ×2.34 = 2 decimal places = 4 decimal places
Multiply the numbers and add.	4.79 ×2.34 1916 1437 +958
Place the decimal point in the answer by counting from the <u>right</u> the same number of decimal places in the problem.	11.2086 = 4 decimal places

48. PLACING THE DECIMAL POINT

DETECTION This may be a special skill need of students who:

- Reverse the position of digits in the answer
- Omit or misplace the decimal point
- Place the answer too far to the left or right of the decimal
- Multiply but use the addition rule to place the decimal

Description. Placement of the decimal point in the multiplication of decimals is a structural process. Problems can also be due to the lack of place-value understanding. It is necessary to differentiate between these two types of problems. The first will be dealt with in this section; the second was dealt with in Special Need 46. The decimal point must be placed with regard to the total decimal points contained in both numbers of the problem. That is, if you are multiplying two numbers, both containing decimal parts, the total number of places to the right of the decimal point in both numbers will set the place for the decimal point in the answer. Normally, the problems will be arranged with the decimal points lined up. This can actually confuse students because the multiplication problem then closely resembles addition and subtraction problems. Students will need to learn to apply the correct set of steps based on which sign precedes the problem, rather than the set-up of the problem alone.

Causation. As mentioned earlier, this type of difficulty can be either conceptual or the result of learning a faulty process for decimal placement. Students who have learned a defective process for decimal placement have probably practiced that process for quite some time and will be consistent in their particular approach. Students may line up the numbers incorrectly and thus place the decimal point incorrectly in either the multiplier or the multiplicand. This type of mistake will cause the students to become even more confused.

Implications. In general, students who exhibit placement problems that result from place-value confusion will need intense experience with manipulatives to grasp the meaning of the parts and whole. This was dealt with in Special Need 46. It will be important to bridge the gap between the newly developed place-value understanding and the correct structural procedure for decimal placement. Do not allow students to develop a defective algorithm after they have achieved this place-value understanding. To develop a correct procedure for decimal placement in the multiplication of decimals, the students could use a set of rules or a flowchart summarizing that set of rules.

CORRECTION Modify these strategies according to students' learning needs.

1. *Concept Review.* Review the exercises in Special Need 46 with the students. Make sure they have a firm grasp of the concepts of the whole and its parts.
2. *Multiplication of Decimals.* Use the following set of steps to help the students develop the sequence to correctly place the decimal in the multiplication of decimals. Practice using these steps with the problems below.

   a. Make sure that all the columns are lined up in the proper order.
   b. Multiply the numbers in each column, using regrouping when necessary.
   c. Count the places to the right of the decimal point for both numbers.
   d. Add the number of places to the right of the decimal in the multiplier and the number of places to the right of the decimal in the multiplicand.
   e. Start at the far-right-hand side of the product and count the same number of places as the result of step d; place the decimal point.

1)	2)	3)	4)
5.5	15.32	33.5	42.932
x 2.1	x 4.32	x 13.2	x 4.654

3. *Extended Practice.* Once you have modeled the appropriate steps for the students, have them work in groups of 2, creating problems and answering them. This can be expanded into a class game of Placing the Point, where teams challenge each other with their problems.

49. DEMONSTRATING PLACE-VALUE UNDERSTANDING

DETECTION This may be a special skill need of students who:

- Leave out the decimal point
- Divide numbers without regard to the column values
- Record their answers without regard to decimal values
- Cannot show the relationship of the parts to the whole

Description. The concept of place value for the division of decimals is a direct outgrowth of that for the division of whole numbers. Students must extend their understanding of the value of each column to the right as well as to the left of the decimal point. Students who had difficulty grasping the concept of place value for the division of whole numbers will almost surely have difficulty understanding place value for the division of decimals. As with fractions, students must see the relationship of the parts and the whole. The advantage to working with decimals is that the entire system is set up in base ten. With fractions, the base is constantly changing as a function of the denominator; with decimals, it remains constant and can be related to the whole-number value. One concept that is particularly difficult for some students is that just because a number has more digits, it may not be a larger number. With whole numbers, if a number has an extra digit (432 instead of 32), the number is automatically greater. With decimals, that is not necessarily the case; it is the value of the tenths place, not the length of the decimal, that most often determines relative value.

Causation. The most obvious cause for difficulty with place-value concepts in the division of decimals is difficulty with place-value concepts in the division of whole numbers. Students also often have difficulty visualizing tenths or hundredths of objects. The concept of breaking down whole objects into partial states will be difficult for students with special learning needs.

Implications. The previous statement does not mean that students with special learning needs cannot learn the concepts of division of decimals. If that were the case, they would not be able to use decimals at all. But there is no practical way to use numbers if you do not understand their function and basic concepts. Thus, special students will need more experience using decimal parts.

CORRECTION Modify these strategies according to students' learning needs.

1. *Fraction Review.* Review with students the relationship of the parts and whole they learned in fractions. Ask them to name items that can be divided into equal parts. See how many the class can name.
2. *Decimal Chart.* Put a decimal place-value chart on the board.

    Example: 4.6

ones	.	tenths
4	.	6

    Read a list of numbers to the class and after modeling the first 2–3, have individual students fill in the place-value chart.
3. *Counting Graph.* Using graph paper, give each student a square with 100 blocks in it, 10 x 10. On the reverse side, they should write "1 piece." Have them turn over the paper and explain that something can be broken down into 100 pieces as easily as 10 pieces and that these parts are called hundredths. Have the students cover 5 hundredths. Next have them cover 10 hundredths, then 15 hundredths, etc. Then go back and write those numbers on the board: .05 = 5 hundredths, etc. Write several additional hundredths values on the board and have students cover the appropriate amounts.
4. *Decimal Chart II.* Put a place-value chart on the board again for tens, ones, tenths, and hundredths. Read a series of numbers containing the above parts and have the students fill in the parts.

    Example: 29.45

    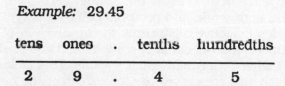

tens	ones	.	tenths	hundredths
2	9	.	4	5

5. *Graph-Paper Division.* Using the graph paper, have students cover 15 of the hundredths. Have them group these 15 hundredths into 3 equal groups. Have them count how many are in each group (5). Explain that the division of decimals works on the same premise as the division of whole numbers, except that you are dividing parts of a whole instead of wholes. Practice the above with many combinations of tenths and hundredths values divided into various groups.

50. APPLYING ALGORITHM SEQUENCE

DETECTION This may be a special skill need of students who:

- Often get events out of order
- Omit steps in a procedure
- Utilize data incorrectly or repeatedly
- Add the divisor to the dividend
- Subtract the divisor from the dividend
- Omit or misplace the decimal point

Description. A sequence of at least 10 skills is involved in the division of decimals. To move from one skill to the next, students need to learn a correct algorithm form. An algorithm is the sequence of events or steps that go together to successfully complete any one skill in the division of decimals. The most common errors made in division of decimals algorithms are the same as those made in division of whole numbers algorithms. In fact, students who demonstrated errors such as those described for Special Needs 28–30 and who did not receive appropriate corrective instruction have a high probability of making the same errors in the division of decimals. In addition to those errors, students frequently omit the decimal point from their answers.

Causation. Students who exhibit this special need are often referred to as sequentially confused learners or as students with sequential learning difficulties. They frequently get steps out of order and also leave steps out of the algorithm. When they arrive at an incorrect answer, they usually cannot find the exact step that caused their answer to be incorrect. These students frequently start one activity and move to another, alternating and never quite completing (at least in a reasonable sequence) any one event. Disorder in schoolwork, homework, and personal behavior is typical.

Implications. Students who develop defective algorithms in the division of decimals will have difficulty with the other decimal operations and will likely develop serious difficulties in higher-order mathematics skills. This lack of organization and structure will also cross over to other areas of the curriculum. Some students do not appear to be highly structured yet are very successful. These students should not be included in the group under discussion, for they have established some alternative method of approaching skills. However, students with defective algorithms and structural weakness need to learn a useful set of steps by which to approach skill acquisition. One of the most powerful means of doing this is the use of flowcharts that depict the proper sequence of events within a skill.

CORRECTION Modify these strategies according to students' learning needs.

1. *Key Label.* Have students make up titles or labels for the algorithm to assist them in retrieving proper sequences for use. As they are labeling the process, they are creating a means to call it back through association.

2. *Problem Structure.* Diagram a sample problem for the students using lines, columns, arrows, or some other tools that will clearly demonstrate a step-by-step procedure.

3. *Algorithm Flowchart:* **Dividing Decimals.** Use the following flowchart to provide students with a visual picture of the total process within which each step fits. Have the students use a place holder (such as a coin) to help organize their movement through the algorithm. Students should use the chart for self-monitoring until the skill is learned.

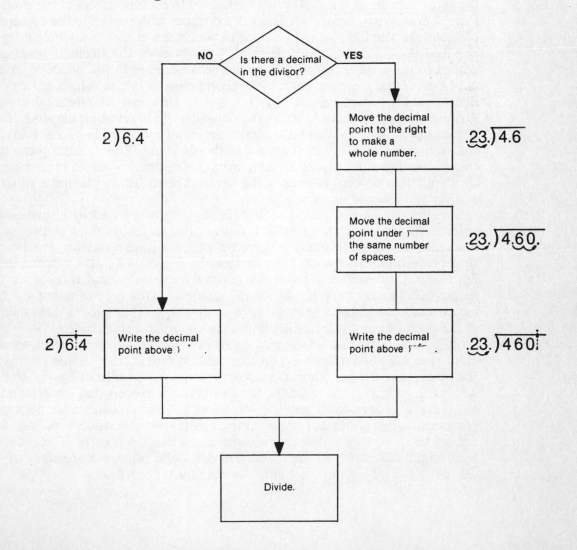

51. PLACING THE DECIMAL POINT

DETECTION This may be a special skill need of students who:

- Reverse the positions of digits in the answer
- Omit the decimal point
- Place the answer too far to the left or right of the decimal

Description. Placement of the decimal point in the division of decimals is a structural process. Problems with placement can also reflect the lack of place-value understanding. It is necessary to differentiate between these two types of problems. The first will be dealt with in this section; the second was dealt with in Special Need 49. The decimal point must be placed with regard to the total number of decimal points contained in both numbers of the problem. There are two broad categories of examples with regard to the division of decimals. The first situation is one in which the divisor is a whole number and the dividend has a decimal. In this case, the decimal point is placed in the answer directly above the decimal point in the dividend; this is of course the simplest case. The second case is one in which both the divisor and the dividend contain decimals. In this case, the decimal point is moved enough spaces to make the divisor a whole number; likewise, the decimal point in the dividend is moved an equal number of spaces, even if that requires the addition of zeroes in the dividend. The decimal point in the quotient is then placed directly over the decimal point in the revised dividend. This process of moving the decimal point before placing it in the answer will confuse many students.

Causation. As mentioned earlier, this type of difficulty can be caused by either conceptual problems or by learning a faulty process for decimal placement. Students who have learned a defective process have probably practiced that process for quite some time and will be consistent in their particular approach. Students may move the decimal incorrectly and thus place it incorrectly in either the divisor or the dividend. This type of mistake will cause students to become even more confused, so that proper movement of the decimal point within the problem is an important consideration.

Implications. In general, students whose placement problems result from place-value confusion will need intense experience dealing with manipulatives to grasp the meaning of the parts and the whole. This was dealt with in Special Need 49. It will be important to bridge the gap between place-value understanding (once it is developed) and the correct structural procedure for decimal placement. After students have achieved this place-value understanding, do not let them develop a defective algorithm. To develop a correct procedure for decimal placement in the division of decimals, provide students with a set of rules or a flowchart that summarizes that set of rules.

CORRECTION Modify these strategies according to students' learning needs.

1. *Concept Review.* Review the exercises in Special Need 49 with the students. Make sure they have a firm grasp of the concepts of the whole and its parts.
2. *Division of Decimals.* Use the following set of steps to help students develop the sequence to correctly place the decimal in the division of decimals. Practice using the flowchart with the problems below.

   a. For problems whose divisor is a whole number:
      1) Put the problem into traditional format. $X\overline{)Y}$
      2) Check that the divisor is a whole number.
      3) Find the decimal point in the dividend.
      4) Place the decimal point in the quotient directly above the decimal point in the dividend.
   b. For problems whose divisor contains a decimal:
      1) Put the problem into traditional format. $X\overline{)Y}$
      2) Check that the divisor has a decimal point.
      3) Move the decimal point to the right, making the divisor a whole number.
      4) Count the number of spaces the decimal point moved in step 3.
      5) Move the decimal to the right the same number of places in the dividend as in step 3.
      6) Place the decimal point in the quotient directly above the decimal point in the revised dividend.

3. *Extended Practice.* Once you have modeled the appropriate steps for students, have them work in groups of 2, creating problems and answering them. This can be expanded into a class game of Placing the Point, where teams challenge each other with their problems.

REFLECTIONS

1. The organization of Part VI suggests different types of problems in the computation of decimal numbers than in other areas of mathematics. Compare and contrast a problem in each area. Are there distinct differences in the DETECTION behaviors and the CORRECTION strategies? Why or why not?

2. Similar observable behaviors are cited for several special skill needs; scan the DETECTION behaviors to locate commonalities across categories. Which particular skills cite the most similar behaviors? Why do you think this is so? Follow a similar procedure to compare CORRECTION strategies across skills.

3. Many of the CORRECTION strategies apply to all students. Justify the selection of the ones presented, adding, deleting, or modifying where you deem necessary.

4. Problems in mathematics tend to assume different proportions according to the student population and the perceptions of individual teachers. Interview a highly skilled regular education teacher to determine his or her perception of the important DETECTION behaviors and CORRECTION strategies for each categorized problem; then discuss detection and correction of any frequent problems that are not mentioned in Chapters 17–20. Follow a similar procedure to interview a veteran special education teacher.

5. Both teaching and learning mathematics are complicated processes. Based on the discussions of special mathematics needs in the previous chapters and your experience, for which difficulties in the computation of decimal numbers do you think regular classroom mathematics instruction is the most difficult to provide? Why? For which is it easiest? Why?

6. The CORRECTIVE PRINCIPLES in Part I are suggested as guides for selecting and modifying the CORRECTION strategies in Part VI. Select a hypothetical special learner; using as guidelines the appropriate CORRECTIVE PRINCIPLES, plan for that learner a modified basal mathematics lesson. Repeat the process for a second special learner. Compare and contrast the two lessons. For the same content, review the lesson script in the teacher's edition of a basal. How do your lessons differ from the ones suggested for most students?

7. Teachers often know how but do not have time to plan special mathematics lessons for special learners. Volunteer your services to plan one or more special mathematics lessons for a special learner in a nearby classroom. Take the mathematics content of your lesson from the basal or other materials currently in use in that school. Use the diagnostic information available from the school and the CORRECTIVE PRINCIPLES to guide the design of your lesson.

8. Teachers also have trouble finding enough time to teach special mathematics lessons. Volunteer to actually teach the mathematics lessons you designed.

9. A number of mathematics and special education textbooks address the mathematics needs of special categories of students. Compare and contrast discussions in these sources with the information in Chapters 17–20:

Ashlock, R. B. (1982). *Error patterns in computation.* Columbus, OH: Charles E. Merrill.

Choate, J. S., Bennett, T. Z., Enright, B. E., Miller, L. J., Poteet, J. A., & Rakes, T. A. (1987). *Assessing and programming basic curriculum skills.* Boston: Allyn and Bacon.

Enright, B. E. (1985). *ENRIGHT computation series* (Books J–M). N. Billerica, MA: Curriculum Associates.

Enright, B. E. (1983). *ENRIGHT diagnostic inventory of basic arithmetic skills.* N. Billerica, MA: Curriculum Associates.

Heddens, J. W. (1984). *Today's mathematics.* Chicago: SRA.

Howell, K. W., & Morehead, M. K. (1987). *Curriculum based evaluation for special and remedial education.* Columbus, OH: Charles E. Merrill.

Johnson, S. W. (1979). *Arithmetic and learning disabilities.* Boston: Allyn and Bacon.

Reisman, F. K. (1982). *A guide to the diagnostic teaching of arithmetic.* Columbus, OH: Charles E. Merrill.

Index

ABOUT THE AUTHOR

BRIAN E. ENRIGHT has almost two decades of experience as an educator, researcher, author, and consultant. He holds the Ed.D. from the University of Alabama and is presently Associate Professor of Education at the University of North Carolina at Charlotte. As an educator he has taught courses in educational assessment, advanced assessment design, prescriptive teaching, and research design. He formerly taught at the elementary, middle, and secondary school levels. His research has focused on the effects of diagnostic/prescriptive instruction, the development of problem solving skills, and the development and norming of criterion-referenced tests.

As an author, he has written over 30 mathematics books, including *The ENRIGHT Diagnostic Inventory of Basic Arithmetic Skills, ENRIGHT Computation Series,* and *SOLVE: Action Problem Solving* (1987). He is co-author of *Assessing and Programming Basic Curriculum Skills* (Allyn and Bacon, 1987) and is also a member of the author team of a major basal mathematics series.

Dr. Enright provides consultant services in the area of mathematics instruction at the national, state, and local levels. He has presented more than 70 papers at international, national, and state conferences. He typically conducts over 50 inservice workshops each year on the topics of diagnostic/prescriptive teaching of mathematics and developing problem solving strategies.

Among his professional activities are service as an associate editor of *Childhood Education, Diagnostique,* and *Behavior Disorders.* He currently serves as a field editor and special column editor for *Teaching Exceptional Children.* He has also served as an officer of several national education boards and is former President of the Council for Educational Diagnostic Services (CEDS).

READER'S REACTION

Dear Reader:

No one knows better than you the special needs of your students or the exact nature of your classroom problems. Your analysis of the extent to which this book meets *your* special needs will help us revise and improve this book, and assist us in developing other books in the *Allyn and Bacon Detecting and Correcting Series*.

Please take a few minutes to respond to the questionnaire on the next page. If you would like to receive a reply to your comments or additional information about the series, indicate this preference in your answer to the last question. Mail the completed form to:

> Joyce S. Choate, Consulting Editor
> Detecting and Correcting Series
> c/o Allyn and Bacon
> 160 Gould Street
> Needham Heights, Massachusetts 02194-2310

Thank you for sharing your special needs and professional concerns.

Sincerely,

Joyce S. Choate

Joyce S. Choate

READER'S REACTION TO

Basic Mathematics: Detecting and Correcting Special Needs

Name: _____   Position: _____

Address: _____   _____

_____   Date: _____

1. How have you used this book?

   ___College Text   ___Inservice Training   ___Teaching Resource

Describe:_____

2. For which purpose(s) do you recommend its use?

3. What do you view as the major strengths of the book?

4. What are its major weaknesses?

5. How could the book be improved?

6. What additional topics should be included in this book?

7. In addition to the books currently included in the *Allyn and Bacon Detecting and Correcting Series*— Basic Mathematics, Classroom Behavior, Language Arts, and Reading— what other books would you recommend developing?

8. Would you like to receive these items?

   ___A reply to your comments

   ___Additional information about this series

Additional Comments:

THANK YOU FOR SHARING YOUR SPECIAL NEEDS AND PROFESSIONAL CONCERNS